On Generalized Growth Rates of Integer Translated Entire and Meromorphic Functions

Authored by

Tanmay Biswas

Research Scientist, Rajbari, Rabindrapally, R. N. Tagore Road
P.O. Krishnagar, P.S. Kotwali, Dist.-Nadia
PIN- 741101, West Bengal, India

&

Chinmay Biswas

Department of Mathematics
Nabadwip Vidyasagar College Nabadwip
Dist.- Nadia, PIN-741302
West Bengal, India

On Generalized Growth rates of Integer Translated Entire and Meromorphic Functions

Authors: Tanmay Biswas and Chinmay Biswas

ISBN (Online): 978-981-5123-61-6

ISBN (Print): 978-981-5123-62-3

ISBN (Paperback): 978-981-5123-63-0

First published in 2023.

Bentham Science Publishers Pte. Ltd.
80 Robinson Road #02-00
Singapore 068898
Singapore
Email: subscriptions@benthamscience.net

BENTHAM
SCIENCE

Contents

PREFACE

The aim of this monograph is to discuss in detail about the growth properties of integer translated entire and meromorphic functions on the basis of their $(p; q; t)$ L-order and $(p; q; t)$ L-type. This book contains six chapters where we step by step elaborate the topic.

Chapter 1 contains the preliminary definitions and notations. In **Chapter 2** and **Chapter 3**, we have derived some results related to $(p; q; t)$ L-th order and $(p; q; t)$ L-th lower order of composite entire and meromorphic functions on the basis of their integer translation. In **Chapter 4**, we have established some relations of integer translated composite entire and meromorphic functions on the basis of their $(p; q; t)$ L-th type and $(p; q; t)$ L-th weak type. **Chapter 5** deals with some results about $(p; q; t)$ L-th order and $(p; q; t)$ L-th type of composite entire and meromorphic functions on the basis of their integer translation. **Chapter 6** is focused on some results about $(p; q; t)$ L-th order and $(p; q; t)$ L-th type of composite entire and meromorphic functions on the basis of their integer translation.

We are thankful to the authors whose publications help us to develop the results of this monograph. We think this monograph will be very helpful for future researchers and students. We are also grateful to Bentham Science Publishers for giving us the opportunity to publish this monograph.

CONSENT FOR PUBLICATION

Not applicable.

CONFLICT OF INTEREST

The authors declare no conflict of interest, financial or otherwise.

Tanmay Biswas
Research Scientist, Rajbari, Rabindrapally, R. N. Tagore Road
P.O. Krishnagar, P.S. Kotwali, Dist.-Nadia
PIN- 741101, West Bengal, India

&

Chinmay Biswas
Department of Mathematics
Nabadwip Vidyasagar College Nabadwip
Dist.- Nadia, PIN-741302
West Bengal, India

ACKNOWLEDGEMENTS

The authors of the present book are very much thankful to the authors of different publications as many new ideas are created from them, and their publications are arranged in the Bibliography. The authors are highly thankful to Bentham Science Publishers for providing the opportunity to publish this book and very much thankful to Ms. Asma Ahmed for her kind correspondence. Authors also express gratefulness to their family members for their continuous help, inspiration, encouragement and sacrifices, lack of which the book cannot be executed. The authors will remain ever grateful to those who helped by giving constructive suggestions so that the work got a good finish. The authors are also responsible for any possible errors and shortcomings, whatever they may be in the book, despite the best attempt to make it immaculate.

(The Authors)

CHAPTER 1

Preliminary Definitions and Notations

Abstract: In this chapter, we discuss the introductory parts connected to the growth of entire and meromorphic functions and some definitions relating to the growth indicators such as order, type, generalized order, generalized type, m-th generalized ${}_pL^*$-order, m-th generalized ${}_pL^*$-type, $(p,q,t)L$-th order, $(p,q,t)L$-th type.

Keywords: Entire function, meromorphic function, generalized order, generalized type, m-th generalized ${}_pL^*$-order, m-th generalized ${}_pL^*$-type, $(p,q,t)L$-th order, $(p,q,t)L$-th type.
Mathematics Subject Classification (2020) : 30D30, 30D35.

1.1 Introduction

Let us consider that the reader is familiar with the fundamental results and the standard notations of Nevanlinna's theory of meromorphic functions, which are available in [1-3]. We also use the standard notations and definitions of the theory of entire functions, which are available in [4, 5]. Some related basic theories of entire and meromorphic functions are briefly discussed in [6, 7], so here we do not repeat those.

Throughout this monograph, we consider that $x \in [0,\infty)$ and $k \in \mathbb{N}$ where $\mathbb{N}$ be the sets of positive integers. We define $\exp^{[k]} x = \exp(\exp^{[k-1]} x)$ and $\log^{[k]} x = \log(\log^{[k-1]} x)$. We also denote $\log^{[0]} x = x$, $\log^{[-1]} x = \exp x$, $\exp^{[0]} x = x$ and $\exp^{[-1]} x = \log x$.

1.2 Preliminary Definitions and Notations

Considering above, the following definitions are relevant and have been frequently used in the monograph.

Tanmay Biswas and Chinmay Biswas

Definition 1.2.1 *The order $\rho(f)$ and the lower order $\lambda(f)$ of an entire function $f(z)$ are defined as*

$$\rho(f) = \limsup_{r \to +\infty} \frac{\log^{[2]} M(r,f)}{\log r} \text{ and } \lambda(f) = \liminf_{r \to +\infty} \frac{\log^{[2]} M(r,f)}{\log r}.$$

For meromorphic $f(z)$,

$$\rho(f) = \limsup_{r \to +\infty} \frac{\log T(r,f)}{\log r} \text{ and } \lambda(f) = \liminf_{r \to +\infty} \frac{\log T(r,f)}{\log r}.$$

Next, to compare the growth of entire or meromorphic functions having the same order, one may give the definitions of type and lower type in the following manner:

Definition 1.2.2 *The type $\sigma(f)$ and the lower type $\overline{\sigma}(f)$ of an entire function $f(z)$ are defined as*

$$\sigma(f) = \limsup_{r \to +\infty} \frac{\log M(r,f)}{r^{\rho(f)}} \text{ and } \overline{\sigma}(f) = \liminf_{r \to +\infty} \frac{\log M(r,f)}{r^{\rho(f)}},$$

where $0 < \rho(f) < \infty$.

If $f(z)$ is meromorphic, then

$$\sigma(f) = \limsup_{r \to +\infty} \frac{T(r,f)}{r^{\rho(f)}} \text{ and } \overline{\sigma}(f) = \liminf_{r \to +\infty} \frac{T(r,f)}{r^{\rho(f)}},$$

where $0 < \rho(f) < \infty$.

It is obvious that $0 \leq \overline{\sigma}(f) \leq \sigma(f) \leq \infty$.

Likewise, to compare the growth of entire or meromorphic functions having the same lower order, one may give the definitions of upper weak type and weak type in the following manner:

Definition 1.2.3 [8] *The upper weak type $\tau(f)$ and the weak type $\overline{\tau}(f)$ of an entire function $f(z)$ of finite positive lower order $\lambda(f)$ are defined by*

$$\tau(f) = \limsup_{r \to +\infty} \frac{\log M(r,f)}{r^{\lambda(f)}} \text{ and } \overline{\tau}(f) = \liminf_{r \to +\infty} \frac{\log M(r,f)}{r^{\lambda(f)}},$$

where $0 < \lambda(f) < \infty$.

If $f(z)$ is meromorphic, then

$$\tau(f) = \limsup_{r \to +\infty} \frac{T(r,f)}{r^{\lambda(f)}} \text{ and } \overline{\tau}(f) = \liminf_{r \to +\infty} \frac{T(r,f)}{r^{\lambda(f)}},$$

where $0 < \lambda(f) < \infty$.

It is obvious that $0 \leq \overline{\tau}(f) \leq \tau(f) \leq \infty$.

Definition 1.2.4 *The hyper order* $\overline{\rho}(f)$ *and the hyper lower order* $\overline{\lambda}(f)$ *of an entire function* $f(z)$ *are defined as*

$$\overline{\rho}(f) = \limsup_{r\to+\infty} \frac{\log^{[3]} M(r,f)}{\log r} \text{ and } \overline{\lambda}(f) = \liminf_{r\to+\infty} \frac{\log^{[3]} M(r,f)}{\log r}.$$

When $f(z)$ *is meromorphic, then*

$$\overline{\rho}(f) = \limsup_{r\to+\infty} \frac{\log^{[2]} T(r,f)}{\log r} \text{ and } \overline{\lambda}(f) = \liminf_{r\to+\infty} \frac{\log^{[2]} T(r,f)}{\log r}$$

holds.

The following two definitions are the natural consequences of the above study:

Definition 1.2.5 *The hyper type* $\widehat{\sigma}(f)$ *and the hyper lower type* $\widehat{\overline{\sigma}}(f)$ *of an entire function* $f(z)$ *are defined as*

$$\widehat{\sigma}(f) = \limsup_{r\to+\infty} \frac{\log^{[2]} M(r,f)}{r^{\overline{\rho}(f)}} \text{ and } \widehat{\overline{\sigma}}(f) = \liminf_{r\to+\infty} \frac{\log^{[2]} M(r,f)}{r^{\overline{\rho}(f)}},$$

where $0 < \overline{\rho}(f) < \infty$.

If $f(z)$ *is meromorphic, then*

$$\widehat{\sigma}(f) = \limsup_{r\to+\infty} \frac{\log T(r,f)}{r^{\overline{\rho}(f)}} \text{ and } \widehat{\overline{\sigma}}(f) = \liminf_{r\to+\infty} \frac{\log T(r,f)}{r^{\overline{\rho}(f)}},$$

where $0 < \overline{\rho}(f) < \infty$.

It is obvious that $0 \leq \widehat{\overline{\sigma}}(f) \leq \widehat{\sigma}(f) \leq \infty$.

Definition 1.2.6 *The hyper upper weak type* $\widehat{\tau}(f)$ *and the hyper weak type* $\widehat{\overline{\tau}}(f)$ *of an entire function* $f(z)$ *of finite positive hyper lower order* $\overline{\lambda}(f)$ *are defined by*

$$\widehat{\tau}(f) = \limsup_{r\to+\infty} \frac{\log^{[2]} M(r,f)}{r^{\overline{\lambda}(f)}} \text{ and } \widehat{\overline{\tau}}(f) = \liminf_{r\to+\infty} \frac{\log^{[2]} M(r,f)}{r^{\overline{\lambda}(f)}},$$

where $0 < \overline{\lambda}(f) < \infty$.

If $f(z)$ *is meromorphic, then*

$$\widehat{\tau}(f) = \limsup_{r\to+\infty} \frac{\log T(r,f)}{r^{\overline{\lambda}(f)}} \text{ and } \widehat{\overline{\tau}}(f) = \liminf_{r\to+\infty} \frac{\log T(r,f)}{r^{\overline{\lambda}(f)}},$$

where $0 < \overline{\lambda}(f) < \infty$.

It is obvious that $0 \leq \widehat{\overline{\tau}}(f) \leq \widehat{\tau}(f) \leq \infty$.

Definition 1.2.7 [9] *Let l be an integer ≥ 1. The generalized order $\rho^{[l]}(f)$ and generalized lower order $\lambda^{[l]}(f)$ of an entire function $f(z)$ are defined as*

$$\rho^{[l]}(f) = \limsup_{r \to +\infty} \frac{\log^{[l+1]} M_f(r)}{\log r} \text{ and } \lambda^{[l]}(f) = \liminf_{r \to +\infty} \frac{\log^{[l+1]} M_f(r)}{\log r}.$$

If $f(z)$ is meromorphic, one can easily verify that

$$\rho^{[l]}(f) = \limsup_{r \to +\infty} \frac{\log^{[l]} T_f(r)}{\log r} \text{ and } \lambda^{[l]}(f) = \liminf_{r \to +\infty} \frac{\log^{[l]} T_f(r)}{\log r}.$$

Remark 1.2.1 *When $l = 1$, Definition 1.2.7 coincides with Definition 1.2.1.*

Next, to compare the growth of entire or meromorphic functions having the same generalized order, one may give the definitions of generalized type and generalized lower type in the following manner:

Definition 1.2.8 *The generalized type $\sigma^{[l]}(f)$ and the generalized lower type $\overline{\sigma}^{[l]}(f)$ of an entire function $f(z)$ are defined as*

$$\sigma^{[l]}(f) = \limsup_{r \to +\infty} \frac{\log^{[l]} M(r, f)}{r^{\rho^{[l]}(f)}} \text{ and } \overline{\sigma}^{[l]}(f) = \liminf_{r \to +\infty} \frac{\log^{[l]} M(r, f)}{r^{\rho^{[l]}(f)}},$$

where $0 < \rho^{[l]}(f) < +\infty$.

If $f(z)$ is meromorphic, then

$$\sigma^{[l]}(f) = \limsup_{r \to +\infty} \frac{\log^{[l-1]} T(r, f)}{r^{\rho^{[l]}(f)}} \text{ and } \overline{\sigma}^{[l]}(f) = \liminf_{r \to +\infty} \frac{\log^{[l-1]} T(r, f)}{r^{\rho^{[l]}(f)}},$$

where $0 < \rho^{[l]}(f) < +\infty$.

It is obvious that $0 \leq \overline{\sigma}^{[l]}(f) \leq \sigma^{[l]}(f) \leq +\infty$.

Likewise, to compare the growth of entire or meromorphic functions having the same generalized lower order, one may give the definitions of generalized upper weak type and generalized weak type in the following manner:

Definition 1.2.9 *The generalized upper weak type $\tau^{[l]}(f)$ and the generalized weak type $\overline{\tau}^{[l]}(f)$ of an entire function $f(z)$ of finite positive generalized lower order $\lambda^{[l]}(f)$ are defined by*

$$\tau^{[l]}(f) = \limsup_{r \to +\infty} \frac{\log^{[l]} M(r, f)}{r^{\lambda^{[l]}(f)}} \text{ and } \overline{\tau}^{[l]}(f) = \liminf_{r \to +\infty} \frac{\log^{[l]} M(r, f)}{r^{\lambda^{[l]}(f)}},$$

where $0 < \lambda^{[l]}(f) < +\infty$.

If $f(z)$ is meromorphic, then

$$\tau^{[l]}(f) = \limsup_{r \to +\infty} \frac{\log^{[l-1]} T(r, f)}{r^{\lambda^{[l]}(f)}} \text{ and } \overline{\tau}^{[l]}(f) = \liminf_{r \to +\infty} \frac{\log^{[l-1]} T(r, f)}{r^{\lambda^{[l]}(f)}},$$

where $0 < \lambda^{[l]}(f) < +\infty$.

It is obvious that $0 \leq \overline{\tau}^{[l]}(f) \leq \tau^{[l]}(f) \leq +\infty$.

Further we recall that Juneja *et al.* [10] defined the (p,q)-th order and (p,q)-th lower order of an entire function, respectively, as follows:

$$\rho^{(p,q)}(f) = \limsup_{r \to +\infty} \frac{\log^{[p+1]} M(r,f)}{\log^{[q]} r} \text{ and } \lambda^{(p,q)}(f) = \liminf_{r \to +\infty} \frac{\log^{[p+1]} M(r,f)}{\log^{[q]} r},$$

where p, q are any two positive integers with $p \geq q$.

If $f(z)$ is meromorphic, one can easily verify that

$$\rho^{(p,q)}(f) = \limsup_{r \to +\infty} \frac{\log^{[p]} T(r,f)}{\log^{[q]} r} \text{ and } \lambda^{(p,q)}(f) = \liminf_{r \to +\infty} \frac{\log^{[p]} T(r,f)}{\log^{[q]} r},$$

where p, q are any two positive integers with $p \geq q$.

For any entire function $f(z)$, using the inequality $T(r,f) \leq \log M(r,f)$ {cf. [11] }, one can easily verify that

$$\begin{matrix} \rho^{(p,q)}(f) \\ \lambda^{(p,q)}(f) \end{matrix} = \lim_{r \to +\infty} \begin{matrix} \sup \\ \inf \end{matrix} \frac{\log^{[p+1]} M(r,f)}{\log^{[q]} r} = \lim_{r \to +\infty} \begin{matrix} \sup \\ \inf \end{matrix} \frac{\log^{[p]} T(r,f)}{\log^{[q]} r},$$

when $p \geq 1$.

Extending the notion (p,q)-th order, recently Shen *et al.* [12] introduced the new concept of $[p,q]$-φ order of entire and meromorphic function where $p \geq q$. Later on, combining the definitions of (p,q)-order and $[p,q]$-φ order, Biswas (*e.g.*, [13]) redefined the (p,q)-order of entire and meromorphic functions without restriction $p \geq q$.

However, the above definition is very useful for measuring the growth of entire and meromorphic functions. If $p = l$ and $q = 1$ then we write $\rho^{(l,1)}(f) = \rho^{(l)}(f)$ and $\lambda^{(l,1)}(f) = \lambda^{(l)}(f)$ where $\rho^{(l)}(f)$ and $\lambda^{(l)}(f)$ are respectively known as generalized order and generalized lower order of entire or meromorphic function $f(z)$. Also for $p = 2$ and $q = 1$, we respectively denote $\rho^{(2,1)}(f)$ and $\lambda^{(2,1)}(f)$ by $\rho(f)$ and $\lambda(f)$, which are classical growth indicators such as order and lower order of entire or meromorphic function $f(z)$.

Further, to compare the growth of entire functions having the same (p,q)-th order, Juneja, Kapoor and Bajpai [14] also introduced the concepts of (p,q)-th type and (p,q)-th lower type in the following manner:

Definition 1.2.10 [14] *The (p,q)-th type $\sigma^{(p,q)}(f)$ and the (p,q)-th lower type $\overline{\sigma}^{(p,q)}(f)$ of an entire function $f(z)$ are defined as*

$$\sigma^{(p,q)}(f) = \limsup_{r \to +\infty} \frac{\log^{[p]} M(r,f)}{\left(\log^{[q-1]} r\right)^{\rho^{(p,q)}(f)}} \text{ and } \overline{\sigma}^{(p,q)}(f) = \liminf_{r \to +\infty} \frac{\log^{[p]} M(r,f)}{\left(\log^{[q-1]} r\right)^{\rho^{(p,q)}(f)}},$$

where $0 < \rho^{(p,q)}(f) < +\infty$.

If $f(z)$ is meromorphic, then

$$\sigma^{(p,q)}(f) = \limsup_{r\to+\infty} \frac{\log^{[p-1]} T(r,f)}{\left(\log^{[q-1]} r\right)^{\rho^{(p,q)}(f)}} \text{ and } \overline{\sigma}^{(p,q)}(f) = \liminf_{r\to+\infty} \frac{\log^{[p-1]} T(r,f)}{\left(\log^{[q-1]} r\right)^{\rho^{(p,q)}(f)}},$$

where $0 < \rho^{(p,q)}(f) < +\infty$.

It is obvious that $0 \le \overline{\sigma}^{(p,q)}(f) \le \sigma^{(p,q)}(f) \le +\infty$.

Likewise, to compare the growth of entire functions having the same (p,q)-th lower order, one can also introduce the concepts of (p,q)-th weak type in the following manner :

Definition 1.2.11 *The (p,q)-th upper weak type $\tau^{(p,q)}(f)$ and the (p,q)-th weak type $\overline{\tau}^{(p,q)}(f)$ of an entire function $f(z)$ of finite positive (p,q)-th lower order $\lambda^{(p,q)}(f)$ are defined by*

$$\tau^{(p,q)}(f) = \limsup_{r\to+\infty} \frac{\log^{[p]} M(r,f)}{\left(\log^{[q-1]} r\right)^{\lambda^{(p,q)}(f)}} \text{ and } \overline{\tau}^{(p,q)}(f) = \liminf_{r\to+\infty} \frac{\log^{[p]} M(r,f)}{\left(\log^{[q-1]} r\right)^{\lambda^{(p,q)}(f)}},$$

where $0 < \lambda^{(p,q)}(f) < +\infty$.

If $f(z)$ is meromorphic, then

$$\tau^{(p,q)}(f) = \limsup_{r\to+\infty} \frac{\log^{[p-1]} T(r,f)}{\left(\log^{[q-1]} r\right)^{\lambda^{(p,q)}(f)}} \text{ and } \overline{\tau}^{(p,q)}(f) = \liminf_{r\to+\infty} \frac{\log^{[p-1]} T(r,f)}{\left(\log^{[q-1]} r\right)^{\lambda^{(p,q)}(f)}},$$

where $0 < \lambda^{(p,q)}(f) < +\infty$.

It is obvious that $0 \le \overline{\tau}^{(p,q)}(f) \le \tau^{(p,q)}(f) \le +\infty$.

Let $L \equiv L(r)$ be a positive continuous function increasing slowly, *i.e.*, $L(ar) \sim L(r)$ as $r \to +\infty$ for every positive constant a. Singh and Barker {cf. [15]} defined it in the following way:

Definition 1.2.12 *A positive continuous function $L(r)$ is called a slowly changing function if for $\epsilon(>0)$,*

$$\frac{1}{k^\epsilon} \le \frac{L(kr)}{L(r)} \le k^\epsilon$$

for $r \ge r(\epsilon)$ and uniformly for $k(\ge 1)$.

Remark 1.2.2 *If $L(r)$ is differentiable, the condition of Definition 1.2.12 is equivalent to*

$$\lim_{r\to+\infty} \frac{rL'(r)}{L(r)} = 0.$$

However, Somasundaram and Thamizharasi {cf. [16]} introduced the notions of L-order and L-lower order for entire functions where $L \equiv L(r)$ is a positive continuous function increasing slowly.

Definition 1.2.13 *The L-order $\rho^L(f)$ and the L-lower order $\lambda^L(f)$ of an entire function $f(z)$ are defined as follows:*

$$\rho^L(f) = \limsup_{r \to +\infty} \frac{\log^{[2]} M(r,f)}{\log[rL(r)]} \text{ and } \lambda^L(f) = \liminf_{r \to +\infty} \frac{\log^{[2]} M(r,f)}{\log[rL(r)]}.$$

When $f(z)$ is meromorphic, then

$$\rho^L(f) = \limsup_{r \to +\infty} \frac{\log T(r,f)}{\log[rL(r)]} \text{ and } \lambda^L(f) = \liminf_{r \to +\infty} \frac{\log T(r,f)}{\log[rL(r)]}.$$

Remark 1.2.3 *For $L(r) \equiv 1$, the definitions of L-order and the L-lower order of entire and meromorphic functions, respectively, reduce to the classical definitions of order and lower order of the same.*

Next, to compare the growth of entire or meromorphic functions having the same L-order, one may give the definitions of L-type and L-lower type in the following manner:

Definition 1.2.14 *The L-type $\sigma^L(f)$ and L-lower type $\overline{\sigma}^L(f)$ of an entire function $f(z)$ having finite positive L-order $\rho^L(f)$ $(0 < \rho^L(f) < \infty)$ are defined as:*

$$\sigma^L(f) = \limsup_{r \to +\infty} \frac{\log M(r,f)}{(rL(r))^{\rho^L(f)}} \text{ and } \overline{\sigma}^L(f) = \liminf_{r \to +\infty} \frac{\log M(r,f)}{(rL(r))^{\rho^L(f)}}.$$

If $f(z)$ is meromorphic, then

$$\sigma^L(f) = \limsup_{r \to +\infty} \frac{T(r,f)}{(rL(r))^{\rho^L(f)}} \text{ and } \overline{\sigma}^L(f) = \liminf_{r \to +\infty} \frac{T(r,f)}{(rL(r))^{\rho^L(f)}}.$$

It is obvious that $0 \leq \overline{\sigma}^L(f) \leq \sigma^L(f) \leq \infty$.

Likewise, to compare the growth of entire or meromorphic functions having the same L-lower order, one can also introduce the concepts of L-upper weak type and L-weak type in the following manner:

Definition 1.2.15 *The L-upper weak type $\tau^L(f)$ and L-weak type $\overline{\tau}^L(f)$ of an entire function $f(z)$ having finite positive L-lower order $\lambda^L(f)$ $(0 < \lambda^L(f) < \infty)$ are defined as:*

$$\tau^L(f) = \limsup_{r \to +\infty} \frac{\log M(r,f)}{(rL(r))^{\lambda^L(f)}} \text{ and } \overline{\tau}^L(f) = \liminf_{r \to +\infty} \frac{\log M(r,f)}{(rL(r))^{\lambda^L(f)}}.$$

If $f(z)$ is meromorphic, then

$$\tau^L(f) = \limsup_{r \to +\infty} \frac{T(r,f)}{(rL(r))^{\lambda^L(f)}} \text{ and } \overline{\tau}^L(f) = \liminf_{r \to +\infty} \frac{T(r,f)}{(rL(r))^{\lambda^L(f)}}.$$

It is obvious that $0 \leq \overline{\tau}^{L}(f) \leq \tau^{L}(f) \leq \infty$.

Definition 1.2.16 *The L^*-order $\rho^{L^*}(f)$ and the L^*-lower order $\lambda^{L^*}(f)$ of an entire function $f(z)$ are defined by*

$$\rho^{L^*}(f) = \limsup_{r\to+\infty}\frac{\log^{[2]} M(r,f)}{\log\left[re^{L(r)}\right]} \text{ and } \lambda^{L^*}(f) = \liminf_{r\to+\infty}\frac{\log^{[2]} M(r,f)}{\log\left[re^{L(r)}\right]}.$$

When $f(z)$ is meromorphic, then

$$\rho^{L^*}(f) = \limsup_{r\to+\infty}\frac{\log T(r,f)}{\log\left[re^{L(r)}\right]} \text{ and } \lambda^{L^*}(f) = \liminf_{r\to+\infty}\frac{\log T(r,f)}{\log\left[re^{L(r)}\right]}.$$

Using the inequality $T(r,f) \leq \log M(r,f)$ {cf. [11] }, one can easily verify that

$$\rho^{L^*}(f) = \limsup_{r\to+\infty}\frac{\log^{[2]} M(r,f)}{\log\left[re^{L(r)}\right]} = \limsup_{r\to+\infty}\frac{\log T(r,f)}{\log\left[re^{L(r)}\right]}$$

and

$$\lambda^{L^*}(f) = \liminf_{r\to+\infty}\frac{\log^{[2]} M(r,f)}{\log\left[re^{L(r)}\right]} = \liminf_{r\to+\infty}\frac{\log T(r,f)}{\log\left[re^{L(r)}\right]}$$

when $e^{L(r)}$ is a positive continuous function increasing slowly, *i.e.*, $e^{L(ar)} \sim e^{L(r)}$ as $r \to +\infty$ for every positive constant a.

Now, for the development of such growth indicators, one may introduce L^*-type and L^*-weak type in the following way:

Definition 1.2.17 *The L^*-type $\sigma^{L^*}(f)$ and L^*-lower type $\overline{\sigma}^{L^*}(f)$ of an entire function $f(z)$ having finite positive L^*-order $\rho^{L^*}(f)$ $(0 < \rho^{L^*}(f) < \infty)$ are defined as:*

$$\sigma^{L^*}(f) = \limsup_{r\to+\infty}\frac{\log M(r,f)}{(re^{L(r)})^{\rho^{L^*}(f)}} \text{ and } \overline{\sigma}^{L^*}(f) = \liminf_{r\to+\infty}\frac{\log M(r,f)}{(re^{L(r)})^{\rho^{L^*}(f)}}.$$

If $f(z)$ is meromorphic, then

$$\sigma^{L^*}(f) = \limsup_{r\to+\infty}\frac{T(r,f)}{(re^{L(r)})^{\rho^{L^*}(f)}} \text{ and } \overline{\sigma}^{L^*}(f) = \liminf_{r\to+\infty}\frac{T(r,f)}{(re^{L(r)})^{\rho^{L^*}(f)}}.$$

It is obvious that $0 \leq \overline{\sigma}^{L^*}(f) \leq \sigma^{L^*}(f) \leq \infty$.

Definition 1.2.18 *The L^*-upper weak type $\tau^{L^*}(f)$ and L^*-weak type $\overline{\tau}^{L^*}(f)$ of an entire function $f(z)$ having finite positive L^*-lower order $\lambda^{L^*}(f)$ $(0 < \lambda^{L^*}(f) < \infty)$ are defined as:*

$$\tau^{L^*}(f) = \limsup_{r\to+\infty}\frac{\log M(r,f)}{(re^{L(r)})^{\lambda^{L^*}(f)}} \text{ and } \overline{\tau}^{L^*}(f) = \liminf_{r\to+\infty}\frac{\log M(r,f)}{(re^{L(r)})^{\lambda^{L^*}(f)}}.$$

If $f(z)$ is meromorphic, then

$$\tau^{L^*}(f) = \limsup_{r\to+\infty}\frac{T(r,f)}{(re^{L(r)})^{\lambda^{L^*}(f)}} \text{ and } \overline{\tau}^{L^*}(f) = \liminf_{r\to+\infty}\frac{T(r,f)}{(re^{L(r)})^{\lambda^{L^*}(f)}}.$$

It is obvious that $0 \leq \overline{\tau}^{L^*}(f) \leq \tau^{L^*}(f) \leq \infty$.

Further, following the above definitions, for any two positive integers m and p, Datta and Biswas [1] introduced the following:

Definition 1.2.19 *The m-th generalized $_pL^*$-order with rate p denoted by ${}^{(m)}_{(p)}\rho^{L^*}(f)$ and the m-th generalized $_pL^*$-lower order with rate p denoted as ${}^{(m)}_{(p)}\lambda^{L^*}(f)$ of an entire function $f(z)$ are defined in the following way:*

$$ {}^{(m)}_{(p)}\rho^{L^*}(f) = \limsup_{r\to+\infty} \frac{\log^{[m+1]} M(r,f)}{\log\left[r \exp^{[p]} L(r)\right]} \text{ and } {}^{(m)}_{(p)}\lambda^{L^*}(f) = \liminf_{r\to+\infty} \frac{\log^{[m+1]} M(r,f)}{\log\left[r \exp^{[p]} L(r)\right]}, $$

where both m and p are positive integers.
When $f(z)$ is meromorphic, then

$$ {}^{(m)}_{(p)}\rho^{L^*}(f) = \limsup_{r\to+\infty} \frac{\log^{[m]} T(r,f)}{\log\left[r \exp^{[p]} L(r)\right]} \text{ and } {}^{(m)}_{(p)}\lambda^{L^*}(f) = \liminf_{r\to+\infty} \frac{\log^{[m]} T(r,f)}{\log\left[r \exp^{[p]} L(r)\right]}, $$

Using the inequality $T(r,f) \leq \log M(r,f)$ {cf. [11] }, one can easily verify that

$$ {}^{(m)}_{(p)}\rho^{L^*}(f) = \limsup_{r\to+\infty} \frac{\log^{[m+1]} M(r,f)}{\log\left[r \exp^{[p]} L(r)\right]} = \limsup_{r\to+\infty} \frac{\log^{[m]} T(r,f)}{\log\left[r \exp^{[p]} L(r)\right]} $$

and

$$ {}^{(m)}_{(p)}\lambda^{L^*}(f) = \liminf_{r\to+\infty} \frac{\log^{[m+1]} M(r,f)}{\log\left[r \exp^{[p]} L(r)\right]} = \liminf_{r\to+\infty} \frac{\log^{[m]} T(r,f)}{\log\left[r \exp^{[p]} L(r)\right]} $$

when $\exp^{[p]} L(r)$ is a positive continuous function increasing slowly, *i.e.*, $\exp^{[p]} L(ar) \sim \exp^{[p]} L(r)$ as $r \to +\infty$ for every positive constant a.

Now, for the development of such growth indicators, one may introduce generalized $_pL^*$-type with rate p and generalized $_pL^*$-weak type with rate p in the following way:

Definition 1.2.20 *The m-th generalized $_pL^*$-type with rate p denoted by ${}^{(m)}_{(p)}\sigma^{L^*}(f)$ and m-th generalized $_pL^*$-lower type with rate p of an entire function $f(z)$ denoted by ${}^{(m)}_{(p)}\overline{\sigma}^{L^*}(f)$ are respectively defined as follows:*

$$ {}^{(m)}_{(p)}\sigma^{L^*}(f) = \limsup_{r\to\infty} \frac{\log^{[m]} M(r,f)}{\left[r \exp^{[p]} L(r)\right]^{{}^{(m)}_{(p)}\rho^{L^*}(f)}} \text{ and} $$

$$ {}^{(m)}_{(p)}\overline{\sigma}^{L^*}(f) = \liminf_{r\to\infty} \frac{\log^{[m]} M(r,f)}{\left[r \exp^{[p]} L(r)\right]^{{}^{(m)}_{(p)}\rho^{L^*}(f)}}, \quad 0 < {}^{(m)}_{(p)}\rho^{L^*}(f) < +\infty, $$

where m and p are any two positive integers.
For meromorphic $f(z)$,

$$ {}^{(m)}_{(p)}\sigma^{L^*}(f) = \limsup_{r\to\infty} \frac{\log^{[m-1]} T(r,f)}{\left[r \exp^{[p]} L(r)\right]^{{}^{(m)}_{(p)}\rho^{L^*}(f)}} \text{ and} $$

$$ {}^{(m)}_{(p)}\overline{\sigma}^{L^*}(f) = \liminf_{r\to\infty} \frac{\log^{[m-1]} T(r,f)}{\left[r \exp^{[p]} L(r)\right]^{{}^{(m)}_{(p)}\rho^{L^*}(f)}}, \quad 0 < {}^{(m)}_{(p)}\rho^{L^*}(f) < +\infty, $$

where both m and p are positive integers.
It is obvious that $0 \leq {}^{(m)}_{(p)}\overline{\sigma}^{L^*}(f) \leq {}^{(m)}_{(p)}\sigma^{L^*}(f) \leq \infty$.

Remark 1.2.4 *If $m = p = 1$, then Definition 1.2.20 becomes the classical one. If $p = 1$ and m is any positive integer, we get the definition of generalized L^*-type $\sigma^{[m]L^*}(f)$ (respectively generalized L^*-lower type $\overline{\sigma}^{[m]L^*}(f)$) and if $m = 1$ and p is any positive integer, then ${}^{(1)}_{(p)}\sigma^{L^*}(f) = {}_{(p)}\sigma^{L^*}(f)$ and ${}^{(1)}_{(p)}\overline{\sigma}^{L^*}(f) = {}_{(p)}\overline{\sigma}^{L^*}(f)$ are respectively called as ${}_pL^*$-type with rate p and ${}_pL^*$-lower type with rate p of an entire or meromorphic function $f(z)$.*

Definition 1.2.21 *The m-th generalized ${}_pL^*$-upper weak type with rate p denoted by ${}^{(m)}_{(p)}\tau^{L^*}(f)$ and m-th generalized ${}_pL^*$-weak type with rate p denoted by ${}^{(m)}_{(p)}\overline{\tau}^{L^*}(f)$ of an entire function $f(z)$ having finite m-th generalized ${}_pL^*$-lower order ${}^{(m)}_{(p)}\lambda^{L^*}(f)$ $(0 < {}^{(m)}_{(p)}\lambda^{L^*}(f) < +\infty)$ are defined as:*

$$
{}^{(m)}_{(p)}\tau^{L^*}(f) = \limsup_{r\to\infty} \frac{\log^{[m]} M(r,f)}{\left[r \exp^{[p]} L(r)\right]^{{}^{(m)}_{(p)}\lambda^{L^*}(f)}} \quad \text{and}
$$

$$
{}^{(m)}_{(p)}\overline{\tau}^{L^*}(f) = \liminf_{r\to\infty} \frac{\log^{[m]} M(r,f)}{\left[r \exp^{[p]} L(r)\right]^{{}^{(m)}_{(p)}\lambda^{L^*}(f)}}, \quad 0 < {}^{(m)}_{(p)}\lambda^{L^*}(f) < +\infty,
$$

where m and p are any two positive integers.
For meromorphic $f(z)$,

$$
{}^{(m)}_{(p)}\tau^{L^*}(f) = \limsup_{r\to\infty} \frac{\log^{[m-1]} T(r,f)}{\left[r \exp^{[p]} L(r)\right]^{{}^{(m)}_{(p)}\lambda^{L^*}(f)}} \quad \text{and}
$$

$$
{}^{(m)}_{(p)}\overline{\tau}^{L^*}(f) = \liminf_{r\to\infty} \frac{\log^{[m-1]} T(r,f)}{\left[r \exp^{[p]} L(r)\right]^{{}^{(m)}_{(p)}\lambda^{L^*}(f)}}, \quad 0 < {}^{(m)}_{(p)}\lambda^{L^*}(f) < +\infty,
$$

where both m and p are positive integers.
It is obvious that $0 \leq {}^{(m)}_{(p)}\overline{\tau}^{L^*}(f) \leq {}^{(m)}_{(p)}\tau^{L^*}(f) \leq \infty$.

Remark 1.2.5 *Particularly, when $p = 1$ and m is any positive integer, then ${}^{(m)}_{(1)}\tau^{L^*}(f) = \tau^{[m]L^*}(f)$ (respectively ${}^{(m)}_{(1)}\overline{\tau}^{L^*}(f) = \overline{\tau}^{[m]L^*}(f)$) and $m = 1$ and p is any positive integer, then ${}^{(1)}_{(p)}\tau^{L^*}(f) = {}_{(p)}\tau^{L^*}_{f^{L^*}(f)}$ (respectively ${}^{(1)}_{(p)}\overline{\tau}^{L^*}(f) = {}_{(p)}\overline{\tau}^{L^*}(f)$). Clearly ${}^{(1)}_{(1)}\tau^{L^*}(f) = \tau^{L^*}(f)$ (respectively ${}^{(1)}_{(1)}\overline{\tau}^{L^*}(f) = \overline{\tau}^{L^*}(f)$)*

In this connection, we just recall the following definition due to Biswas [17, 18]:

Definition 1.2.22 *The $(p,q,t)L$-th order denoted by $\rho^{(p,q,t)L}(f)$ and the $(p,q,t)L$-th lower order denoted as $\lambda^{(p,q,t)L}(f)$ of an entire function $f(z)$ are defined in the following way:*

$$\rho^{(p,q,t)L}(f) = \limsup_{r\to+\infty} \frac{\log^{[p+1]} M(r,f)}{\log^{[q]} r + \exp^{[t]} L(r)} \quad \text{and}$$
$$\lambda^{(p,q,t)L}(f) = \liminf_{r\to+\infty} \frac{\log^{[p+1]} M(r,f)}{\log^{[q]} r + \exp^{[t]} L(r)},$$

where p, q are any two positive integers and $t \in \mathbb{N} \cup \{-1,0\}$.
If $f(z)$ is a meromorphic function, then

$$\rho^{(p,q,t)L}(f) = \limsup_{r\to+\infty} \frac{\log^{[p]} T(r,f)}{\log^{[q]} r + \exp^{[t]} L(r)} \quad \text{and}$$
$$\lambda^{(p,q,t)L}(f) = \liminf_{r\to+\infty} \frac{\log^{[p]} T(r,f)}{\log^{[q]} r + \exp^{[t]} L(r)},$$

where p, q are any two positive integers and $t \in \mathbb{N} \cup \{-1,0\}$.

Using the inequality $T(r,f) \leq \log M(r,f)$ {cf. [11] }, one can easily verify that

$$\rho^{(p,q,t)L}(f) = \limsup_{r\to+\infty} \frac{\log^{[p+1]} M(r,f)}{\log^{[q]} r + \exp^{[t]} L(r)} = \limsup_{r\to+\infty} \frac{\log^{[p]} T(r,f)}{\log^{[q]} r + \exp^{[t]} L(r)}$$

and

$$\lambda^{(p,q,t)L}(f) = \liminf_{r\to+\infty} \frac{\log^{[p+1]} M(r,f)}{\log^{[q]} r + \exp^{[t]} L(r)} = \liminf_{r\to+\infty} \frac{\log^{[p]} T(r,f)}{\log^{[q]} r + \exp^{[t]} L(r)}$$

when $\exp^{[t]} L(r)$ is a positive continuous function increasing slowly, *i.e.*, $\exp^{[t]} L(ar) \sim \exp^{[t]} L(r)$ as $r \to +\infty$ for every positive constant a. Throughout the present book we always consider that $\exp^{[t]} L(r)$ is a positive continuous function increasing slowly.

In order to compare the relative growth of two entire functions having same nonzero finite $(p,q,t)L$-th order, one may introduce the definition of $(p,q,t)L$-th type (respectively $(p,q,t)L$-th lower type) of entire functions having finite positive finite $(p,q,t)L$-th order in the following manner:

Definition 1.2.23 *The $(p,q,t)L$-th type denoted by $\sigma^{(p,q,t)L}(f)$and $(p,q,t)L$-th lower type of an entire function $f(z)$ denoted by $\overline{\sigma}^{(p,q,t)L}(f)$ are respectively defined as follows:*

$$\sigma^{(p,q,t)L}(f) = \limsup_{r\to+\infty} \frac{\log^{[p]} M(r,f)}{\left[\log^{[q-1]} r \cdot \exp^{[t+1]} L(r)\right]^{\rho^{(p,q,t)L}(f)}} \quad \text{and}$$

$$\overline{\sigma}^{(p,q,t)L}(f) = \liminf_{r\to+\infty} \frac{\log^{[p]} M(r,f)}{\left[\log^{[q-1]} r \cdot \exp^{[t+1]} L(r)\right]^{\rho^{(p,q,t)L}(f)}}, \quad 0 < \rho^{(p,q,t)L}(f) < +\infty,$$

where p, q are any two positive integers and $t \in \mathbb{N} \cup \{-1, 0\}$.
For meromorphic $f(z)$,

$$\sigma^{(p,q,t)L}(f) = \limsup_{r \to +\infty} \frac{\log^{[p-1]} T(r,f)}{\left[\log^{[q-1]} r \cdot \exp^{[t+1]} L(r)\right]^{\rho^{(p,q,t)L}(f)}} \text{ and}$$

$$\overline{\sigma}^{(p,q,t)L}(f) = \liminf_{r \to +\infty} \frac{\log^{[p-1]} T(r,f)}{\left[\log^{[q-1]} r \cdot \exp^{[t+1]} L(r)\right]^{\rho^{(p,q,t)L}(f)}}, \quad 0 < \rho^{(p,q,t)L}(f) < +\infty,$$

where p, q are any two positive integers and $t \in \mathbb{N} \cup \{-1, 0\}$.
It is obvious that $0 \leq \overline{\sigma}^{(p,q,t)L}(f) \leq \sigma^{(p,q,t)L}(f) \leq \infty$.

Analogously in order to determine the relative growth of two entire functions having the same non zero finite $(p, q, t)L$-th lower order, one may introduce the definition of $(p, q, t)L$-th weak type of entire functions having finite positive $(p, q, t)L$-th lower order in the following way:

Definition 1.2.24 *The* $(p, q, t)L$*-th upper weak type denoted by* $\tau^{(p,q,t)L}(f)$ *and* $(p, q, t)L$*-th weak type denoted by* $\overline{\tau}^{(p,q,t)L}(f)$ *of an entire function* $f(z)$ *having finite* $(p, q, t)L$*-th lower order* $\lambda^{(p,q,t)L}(f)$ $(0 < \lambda^{(p,q,t)L}(f) < +\infty)$ *are defined as:*

$$\tau^{(p,q,t)L}(f) = \limsup_{r \to \infty} \frac{\log^{[p]} M(r,f)}{\left[\log^{[q-1]} r \cdot \exp^{[t+1]} L(r)\right]^{\lambda^{(p,q,t)L}(f)}} \text{ and}$$

$$\overline{\tau}^{(p,q,t)L}(f) = \liminf_{r \to \infty} \frac{\log^{[p]} M(r,f)}{\left[\log^{[q-1]} r \cdot \exp^{[t+1]} L(r)\right]^{\lambda^{(p,q,t)L}(f)}}, \quad 0 < \lambda^{(p,q,t)L}(f) < +\infty,$$

where p, q are any two positive integers and $t \in \mathbb{N} \cup \{-1, 0\}$.
For meromorphic $f(z)$,

$$\tau^{(p,q,t)L}(f) = \limsup_{r \to \infty} \frac{\log^{[p-1]} T(r,f)}{\left[\log^{[q-1]} r \cdot \exp^{[t+1]} L(r)\right]^{\lambda^{(p,q,t)L}(f)}} \text{ and}$$

$$\overline{\tau}^{(p,q,t)L}(f) = \liminf_{r \to \infty} \frac{\log^{[p-1]} T(r,f)}{\left[\log^{[q-1]} r \cdot \exp^{[t+1]} L(r)\right]^{\lambda^{(p,q,t)L}(f)}}, \quad 0 < \lambda^{(p,q,t)L}(f) < +\infty,$$

where p, q are any two positive integers and $t \in \mathbb{N} \cup \{-1, 0\}$.
It is obvious that $0 \leq \overline{\tau}^{(p,q,t)L}(f) \leq \tau^{(p,q,t)L}(f) \leq \infty$.

1.3 Concluding Remark

During the past decades, several authors {cf. [1, 6-8, 13, 17-41]} made closed investigations on the growth properties of composite entire and meromorphic functions using different growth indicators. Using the concepts of $(p,q,t)L$-th order and $(p,q,t)L$-th type, in the present book we wish to establish some further growth properties of composite entire and meromorphic functions on the basis of their integer translation. In fact, in the present book, we wish to extend the works presented in [6, 7]. Further, we assume that throughout the present book, l, p, q, p_1, q_1, p_2, q_2, m and n always denote positive integers and $t \in \mathbb{N} \cup \{-1,0\}$.

Considering this, in the next chapter, we wish to establish some related growth analysis of the composition of integer translated entire and meromorphic functions using $(p,q,t)L$-th order and $(p,q,t)L$-th lower order.

References

[1] S. K. Datta and A. Biswas, "Some generalized growth properties of composite entire functions involving their maximum terms on the basis of slowly changing functions", *Int. Journal of Math. Analysis*, Vol. 5, No.22, pp. 1093-1101, 2011.

[2] I. Laine, *Nevanlinna Theory and Complex Differential Equations*, Berlin: De Gruyter, 1993.

[3] L. Yang, *Value distribution theory*, Berlin: Springer-Verlag, 1993.

[4] E.C. Titchmarsh, *The theory of functions* , 2nd ed., Oxford: Oxford University Press, 1968.

[5] G. Valiron, *Lectures on the general theory of integral functions*, New York: Chelsea Publishing Company, 1949.

[6] T. Biswas and C. Biswas, *Integer translation and growth of entire and meromorphic functions*, Republic of Moldova Europe: GlobeEdit a trademark of Dodo Books Indian Ocean Ltd., member of the OmniScriptum S.R.L. Publishing group, 2021.

[7] T. Biswas, C. Biswas and R. Biswas, *Growth of Integer Translated Composite Entire and Meromorphic Functions*, Republic of Moldova Europe: LAP Lambert Academic Publishing, a trademark of Dodo Books Indian Ocean Ltd., member of the OmniScriptum S.R.L. Publishing group, 2022.

[8] S. K. Datta and A. Jha, "On the weak type of meromorphic functions", *Int. Math. Forum*, Vol. 4, No. 12, pp. 569-579, 2009.

[9] D. Sato, "On the rate of growth of entire functions of fast growth", *Bull. Amer. Math. Soc.*, Vol. 69, pp. 411-414, 1963.

[10] O. P. Juneja, G. P. Kapoor and S. K. Bajpai, "On the (p,q)-order and lower (p,q)-order of an entire function", *J. Reine Angew. Math.*, Vol. 282, pp. 53-67, 1976.

[11] W. K. Hayman, *Meromorphic Functions*, Oxford: The Clarendon Press, 1964.

[12] X. Shen, J. Tu and H. Y. Xu, "Complex oscillation of a second-order linear differential equation with entire coefficients of $[p,q]$-φ order", *Adv. Difference Equ.*, 2014(1), 200, 14 pages, 2014.

[13] T. Biswas, "On some inequalities concerning relative (p,q)-φ type and relative (p,q)-φ weak type of entire or meromorphic functions with respect to an entire function", *J. Class. Anal.*, Vol. 13, No 2, pp. 107-122, 2018.

[14] O. P. Juneja, G. P. Kapoor and S. K. Bajpai, "On the (p,q)-type and lower (p,q)-type of an entire function", *J. Reine Angew. Math.*, Vol. 290, pp. 180-190, 1977.

[15] S.K. Singh and G.P. Barker, "Slowly changing functions and their applications", *Indian J. Math.*, Vol. 19, No. 1, pp. 1-6, 1977.

[16] D. Somasundaram and R.Thamizharasi, "A note on the entire functions of L-bounded index and L-type", *Indian J. Pure Appl. Math.*, Vol. 19 , No.3, pp.284-293, March 1988.

[17] T. Biswas, "On some growth analysis of entire and meromorphic functions in the light of their relative $(p,q,t)L$-th order with respect to another entire function", *An. Univ. Oradea Fasc. Mat.*, Tom XXVI, Issue No.1, pp. 59-80, 2019.

[18] T. Biswas, "Some results relating to sum and product theorems of relative $(p,q,t)L$-th order and relative (p,q,t)L-th type of entire functions", *Korean J. Math.*, Vol. 26, No.2, pp. 215-269, 2018.

[19] T. Biswas, "A note on some growth properties of composite entire and meromorphic functions using their relative (p,q)-th orders", *Electron. J. Math. Anal. Appl.*, Vol. 7, No. 2, pp. 151-167, 2019.

[20] T. Biswas, "Some comparative growth rates of wronskians generated by entire and meromorphic functions on the basis of their relative (p,q)-th type and relative (p,q)-th weak type", *J. Fract. Calc. Appl.*, Vol. 10, No. 2, pp. 147-166, 2019.

[21] T. Biswas, "An extensive study on sum and product theorems of relative (p,q) th order and relative (p,q) th type of meromorphic functions with respect to entire functions", *Electron. J. Math. Anal. Appl.*, Vol. 7, No. 1, pp. 33-64, 2019.

[22] T. Biswas, "Central index based some comparative growth analysis of composite entire functions from the view point of L^*- order", *J. Korean Soc. Math. Educ. Ser. B Pure Appl. Math.*, Vol. 25, No3, pp. 193-201, August, 2018.

[23] T. Biswas, "Growth measurements of entire and meromorphic functions on the basis of their integer translation", *Int. J. Nonlinear Sci.*, Vol. 26, No. 3, pp. 140-149, 2018.

[24] T. Biswas, "On some growth analysis of composite entire and meromorphic functions from the view point of their relative (p,q)-th type and relative (p,q)-th weak type", *Korean J. Math.*, Vol. 26, No. 1, pp. 23-41, 2018.

[25] T. Biswas, "Comparative growth analysis of special type of differential polynomial generated by entire and meromorphic functions on the basis of their (p,q)-th order", *Int. J. of Appl. Math.*, Vol. 31, No. 1, pp. 73-92, 2018.

[26] T. Biswas, "Advancement on the study of growth analysis of differential polynomial and differential monomial in the light of slowly increasing functions", *Carpathian Math. Publ.*, Vol. 10, pp. 31-57, 2018.

[27] T. Biswas, "Comparative growth measurement of differential monomials and differential polynomials depending upon their relative ${}_pL^*$-types and relative ${}_pL^*$-weak types" *Aligarh Bull. Math.*, Vol. 36, pp. 73-94, 2017.

[28] T. Biswas, "Comparative growth analysis of differential monomials and differential polynomials depending on their relative ${}_pL^*$-orders", *J. Chungcheong Math. Soc.*, Vol.31, pp. 103-130, 2018.

[29] N. Bhattacharjee and I. Lahiri, "Growth and value distribution of differential polynomials", *Bull. Math. Soc. Sc. Math. Roumanie Tome*, Vol 39 (87), No.1-4, pp.85-104, 1996.

[30] K.S.Charak, "Value distribution theory of meromorphic functions", *Mathematics News Letter*, Vol.18, No.4, pp. 121 -139, 2009.

[31] S. K. Datta, T. Biswas and S. Ali, "Growth rates of wronskians generated by complex valued functions", *Gulf J. Math.*, Vol. 2, Issue 1, pp. 58-74, 2014.

[32] S. K. Datta and T. Biswas, "Growth of entire functions based on relative order", *Int. J. Pure Appl. Math.*, Vol. 51, No.1, pp. 49-58, 2009.

[33] S. K. Datta, T. Biswas and S. Ali (2013), "Growth analysis of wronskians in terms of slowly changing functions", *J. Complex Anal.*, Vol.2013, Article ID 395067, 9 pages, http://dx.doi.org/10.1155/2013/395067.

[34] S. K.Datta , T.Biswas and S.Ali, "Some growth properties of wronskians using their relative order", J. Class. Anal., Vol. 3, No. 1, pp. 91-99, 2013.

[35] S. K. Datta, T. Biswas and S. Ali, "Some results on wronskians using slowly changing functions", *News Bull. Cal. Math. Soc.*, Vol. 36, No. 7-9, pp. 8-24, 2013.

[36] S. K. Datta, T. Biswas and A. Hoque, "On some growth properties of differential polynomials in the light of relative order", *Ital. J. Pure Appl. Math.*, N. 32, pp. 235-246, 2014.

[37] S.K.Datta, T. Biswas and A. Kar, "Slowly changing function based growth analysis of wronskians generated by meromorphic functions", *Aligarh Bull. Math.*, Vol. 33, No. 1-2, pp. 1-14, 2014.

[38] S.K.Datta, T. Biswas and A. Kar, "On the growths of meromorphic function generated wronskians from the view point of slowly changing functions", *Facta Univ. Ser. Math. Inform.*, Vol. 30, No 2, pp. 169-193, 2015.

[39] S.K.Datta, T. Biswas and A. Kar, "Slowly changing function connected growth properties of wronskians generated by entire and meromorphic functions", *Acta Univ. Sapientiae Mathematica*, Vol. 7, No. 2, pp. 141-166, 2015.

[40] S.K.Datta, T. Biswas and A. Kar, "On the growth analysis of wronskians in the light of some generalized growth indicators", *Commun. Fac. Sci. Univ. Ank. Series A1*, Vol. 64, No. 2, pp. 1-34, 2015.

[41] I. Lahiri and N.R. Bhattacharjee, "Functions of L-bounded index and of non-uniform L-bounded index", *Indian J. Math.*, Vol. 40, No. 1, pp. 43-57, 1998.

CHAPTER 2

$(p, q, t)L$-th Order Oriented Some Growth Analysis of Composite Entire and Meromorphic Functions on the Basis of Their Integer Translation

Abstract: The main objective of this chapter is to investigate some results related to the growth rates of the composition of integer translated entire and meromorphic functions using $(p, q, t)L$-th order and $(p, q, t)L$-th lower order.

Keywords: Integer translated entire function, Integer translated meromorphic function, composition, $(p, q, t)L$-th order, $(p, q, t)L$-th lower order.
Mathematics Subject Classification (2020) : 30D30, 30D35.

2.1 Introduction

Let $f(z)$ be a meromorphic function and $n \in \mathbb{N}$, then the translation of $f(z)$ be denoted by $f(z+n)$. For each $n \in \mathbb{N}$, one may obtain a function with some properties. Let us consider this family by $f_n(z)$ where

$$f_n(z) = \{f(z+n) : n \in \mathbb{N}\}.$$

We recall that if α is a regular point of an analytic function $f(z)$ and if $f(\alpha) = 0$, then α is called a zero of $f(z)$. The point $z = \alpha$ is called a zero of $f(z)$ of multiplicity m (m being a positive integer) if in some neighborhood of α, $f(z)$ can be expanded in a Taylor's series of the form $f(z) = \sum_{n=m}^{\infty} a_n (z-\alpha)^n$ where $a_m = 0$.

It is clear that the number of zeros of $f(z)$ may be changed in a finite region

Tanmay Biswas and Chinmay Biswas

after translation but it remains unaltered in the open complex plane $\mathbb{C}$, *i.e.*,

$$N(r, f(z+n)) = N(r, f) + e_n, \tag{2.1.1}$$

where e_n is a residue term such that $e_n \to 0$ as $r \to +\infty$.

Also

$$m(r, f(z+n)) = \frac{1}{2\pi}\int_0^{2\pi} \log^+ \left|f(re^{i\theta}+n)\right| d\theta$$

$$i.e., m(r, f(z+n)) = m(r, f) + e_n^{'}, \tag{2.1.2}$$

where $e_n^{'}$ (may be distinct from e_n) be such that $e_n^{'} \to 0$ as $r \to +\infty$.

Therefore from (2.1.1) and (2.1.2), one may obtain that

$$N(r, f(z+n)) + m(r, f(z+n)) = N(r, f) + e_n + m(r, f) + e_n^{'}$$
$$i.e., T(r, f(z+n)) = T(r, f) + e_n + e_n^{'}.$$

Now if n varies, then Nevanlinna's Characteristic function for the family $f_n(z)$, where $f_n(z) = \{f(z+n) : n \in \mathbb{N}\}$ for the meromorphic function f is

$$T(r, f_n) = nT(r, f) + \sum_n (e_n + e_n^{'}). \tag{2.1.3}$$

Similarly, one can define a family for each $m \in \mathbb{N}$, $g_m(z) = \{g(z+m) : m \in \mathbb{N}\}$ where $g(z)$ is an entire function. Then the composition $f_n \circ g_m$ is defined.

Let $f_n \circ g_m = h_t$, where h is a meromorphic function and $t \in \mathbb{N}$. So h_t can be expressed as $h_t = \{h(z+t) : t \in \mathbb{N}\}$.

Then by (2.1.3)

$$T(r, h_t) = tT(r, h) + \sum_t (e_t + e_t^{'})$$

where $e_t, e_t^{'} \to 0$ as $r \to +\infty$.

$$i.e., T(r, f_n \circ g_m) = tT(r, f(g)) + \sum_t (e_t + e_t^{'}). \tag{2.1.4}$$

However, in the case of any two meromorphic functions $f(z)$ and $g(z)$, the ratio $\frac{T(r,f)}{T(r,g)}$ as $r \to +\infty$ is called as the growth of $f(z)$ with respect to $g(z)$ in terms of their Nevanlinna's Characteristic functions. Further, the concept of the growth measuring tools such as order and lower order which are conventional in complex analysis and the growth of entire or meromorphic functions can be studied in terms of their orders and lower orders.

Somasundaram and Thamizharasi [1] introduced the notions of L-order and L-lower order for entire functions where $L \equiv L(r)$ is a positive continuous function increasing slowly, i.e., $L(ar) \sim L(r)$ as $r \to +\infty$ for every positive constant "a". The more generalized concepts of L-order and L-lower order of meromorphic functions are $(p,q,t)L$-th order and $(p,q,t)L$-th lower order respectively.

The principal objective of this chapter is to investigate some results related to the growth rates of the composition of integer translated entire and meromorphic functions using $(p,q,t)L$-th order and $(p,q,t)L$-th lower order of entire and meromorphic functions.

2.2 Lemmas

In this section, we present some lemmas which will be needed in the sequel.

Lemma 2.2.1 [2] *Let $f(z)$ be a meromorphic function. If $f_n(z) = f(z+n)$ for $n \in \mathbb{N}$ then*

$$\lim_{r\to+\infty} \frac{T(r,f_n)}{T(r,f)} = n.$$

Lemma 2.2.2 *Let $f(z)$ be a meromorphic function. If $f_n(z) = f(z+n)$ for $n \in \mathbb{N}$ then*

$$\rho^{(p,q,t)L}(f_n) = \rho^{(p,q,t)L}(f) \text{ and } \lambda^{(p,q,t)L}(f_n) = \lambda^{(p,q,t)L}(f).$$

Proof By Lemma 2.2.1 $\lim\limits_{r\to+\infty} \frac{\log^{[p]} T(r,f_n)}{\log^{[p]} T(r,f)}$ exists and is equal to 1.

Now,

$$\begin{aligned}
\rho^{(p,q,t)L}(f_n) &= \limsup_{r\to+\infty} \frac{\log^{[p]} T(r,f_n)}{\log^{[q]} r + \exp^{[t]} L(r)} \\
&= \limsup_{r\to+\infty} \left\{ \frac{\log^{[p]} T(r,f)}{\log^{[q]} r + \exp^{[t]} L(r)} \cdot \frac{\log^{[p]} T(r,f_n)}{\log^{[p]} T(r,f)} \right\} \\
&= \limsup_{r\to+\infty} \frac{\log^{[p]} T(r,f)}{\log^{[q]} r + \exp^{[t]} L(r)} \cdot \lim_{r\to+\infty} \frac{\log^{[p]} T(r,f_n)}{\log^{[p]} T(r,f)} \\
&= \rho^{(p,q,t)L}(f) \cdot 1 \\
&= \rho^{(p,q,t)L}(f).
\end{aligned}$$

In a similar manner, $\lambda^{(p,q,t)L}(f_n) = \lambda^{(p,q,t)L}(f)$.

This proves the lemma.

2.3 Main Results

In this section, we present the main results of the chapter.

Theorem 2.3.1 *Let $f(z)$ be a meromorphic function and $g(z)$ be a non constant entire function such that $0 < \lambda^{(m,q,t)L}(f(g)) \le \rho^{(m,q,t)L}(f(g)) < +\infty$ and $0 < \lambda^{(l,q,t)L}(f) \le$*

$\rho^{(l,q,t)L}(f) < +\infty$. *Also let* f_u *and* g_v *be integer translations of* $f(z)$ *and* $g(z)$, *respectively, for* $u, v \in \mathbb{N}$. *Then*

$$\frac{\lambda^{(m,q,t)L}(f(g))}{\rho^{(l,q,t)L}(f)} \leq \liminf_{r\to+\infty} \frac{\log^{[m]} T(r, f_u(g_v))}{\log^{[l]} T(r, f_u)} \leq \frac{\lambda^{(m,q,t)L}(f(g))}{\lambda^{(l,q,t)L}(f)} \leq$$

$$\limsup_{r\to+\infty} \frac{\log^{[m]} T(r, f_u(g_v))}{\log^{[l]} T(r, f_u)} \leq \frac{\rho^{(m,q,t)L}(f(g))}{\lambda^{(l,q,t)L}(f)}.$$

Proof By the procedure of establishing (2.1.4) we can express

$$T(r, f_u(g_v)) = tT(r, f(g)) + \sum_t (e_t + e_t'),$$

where $e_t, e_t' \to 0$ as $r \to +\infty$. Therefore

$$\lim_{r\to+\infty} \frac{T(r, f_u(g_v))}{T(r, f(g))} = t.$$

So in view of Lemma 2.2.2, we get that

$$\rho^{(m,q,t)L}(f_u(g_v)) = \rho^{(m,q,t)L}(f(g)) \text{ and } \lambda^{(m,q,t)L}(f_u(g_v)) = \lambda^{(m,q,t)L}(f(g)). \quad (2.3.1)$$

Now from the definition of $(m, q, t)L$-th order and $(m, q, t)L$-th lower order we have for arbitrary positive ε and for all large values of r,

$$\log^{[m]} T(r, f_u(g_v)) \geqslant (\lambda^{(m,q,t)L}(f_u(g_v)) - \varepsilon)\left(\log^{[q]} r + \exp^{[t]} L(r)\right)$$

$$i, e, \ \log^{[m]} T(r, f_u(g_v)) \geqslant (\lambda^{(m,q,t)L}(f(g)) - \varepsilon)\left(\log^{[q]} r + \exp^{[t]} L(r)\right) \quad (2.3.2)$$

and

$$\log^{[l]} T(r, f_u) \leq (\rho^{(l,q,t)L}(f_u) + \varepsilon)\left(\log^{[q]} r + \exp^{[t]} L(r)\right)$$

$$i.e., \ \log^{[l]} T(r, f_u) \leq (\rho^{(l,q,t)L}(f) + \varepsilon)\left(\log^{[q]} r + \exp^{[t]} L(r)\right). \quad (2.3.3)$$

Now from (2.3.2) and (2.3.3) it follows for all large values of r,

$$\frac{\log^{[m]} T(r, f_u(g_v))}{\log^{[l]} T(r, f_u)} \geqslant \frac{\lambda^{(m,q,t)L}(f(g)) - \varepsilon}{\rho^{(l,q,t)L}(f) + \varepsilon}.$$

As $\varepsilon(> 0)$ is arbitrary , we obtain that

$$\liminf_{r\to+\infty} \frac{\log^{[m]} T(r, f_u(g_v))}{\log^{[l]} T(r, f_u)} \geqslant \frac{\lambda^{(m,q,t)L}(f(g))}{\rho^{(l,q,t)L}(f)}. \quad (2.3.4)$$

Again for a sequence of values of r tending to infinity,

$$\log^{[m]} T(r, f_u(g_v)) \leq (\lambda^{(m,q,t)L}(f_u(g_v)) + \varepsilon)\left(\log^{[q]} r + \exp^{[t]} L(r)\right)$$

$$i.e., \ \log^{[m]} T(r, f_u(g_v)) \leq (\lambda^{(m,q,t)L}(f(g)) + \varepsilon) \left(\log^{[q]} r + \exp^{[t]} L(r)\right) \quad (2.3.5)$$

and for all large values of r,

$$\log^{[l]} T(r, f_u) \geqslant (\lambda^{(l,q,t)L}(f_u) - \varepsilon) \left(\log^{[q]} r + \exp^{[t]} L(r)\right)$$

$$i.e., \ \log^{[l]} T(r, f_u) \geqslant (\lambda^{(l,q,t)L}(f) - \varepsilon) \left(\log^{[q]} r + \exp^{[t]} L(r)\right). \quad (2.3.6)$$

So combining (2.3.5) and (2.3.6) we get for a sequence of values of r tending to infinity,

$$\frac{\log^{[m]} T(r, f_u(g_v))}{\log^{[l]} T(r, f_u)} \leq \frac{\lambda^{(m,q,t)L}(f(g)) + \varepsilon}{\lambda^{(l,q,t)L}(f) - \varepsilon}.$$

Since $\varepsilon(> 0)$ is arbitrary it follows that

$$\liminf_{r \to +\infty} \frac{\log^{[m]} T(r, f_u(g_v))}{\log^{[l]} T(r, f_u)} \leq \frac{\lambda^{(m,q,t)L}(f(g))}{\lambda^{(l,q,t)L}(f)}. \quad (2.3.7)$$

Also for a sequence of values of r tending to infinity,

$$\log^{[l]} T(r, f_u) \leq (\lambda^{(l,q,t)L}(f_u) + \varepsilon) \left(\log^{[q]} r + \exp^{[t]} L(r)\right)$$

$$i.e., \ \log^{[l]} T(r, f_u) \leq (\lambda^{(l,q,t)L}(f) + \varepsilon) \left(\log^{[q]} r + \exp^{[t]} L(r)\right). \quad (2.3.8)$$

Now from (2.3.2) and (2.3.8) we obtain for a sequence of values of r tending to infinity,

$$\frac{\log^{[m]} T(r, f_u(g_v))}{\log^{[l]} T(r, f_u)} \geqslant \frac{\lambda^{(m,q,t)L}(f(g)) - \varepsilon}{\lambda^{(l,q,t)L}(f) + \varepsilon}.$$

Choosing $\varepsilon(> 0)$ we get from above that

$$\limsup_{r \to +\infty} \frac{\log^{[m]} T(r, f_u(g_v))}{\log^{[l]} T(r, f_u)} \geqslant \frac{\lambda^{(m,q,t)L}(f(g))}{\lambda^{(l,q,t)L}(f)}. \quad (2.3.9)$$

Also for all large values of r,

$$\log^{[m]} T(r, f_u(g_v)) \leq (\rho^{(m,q,t)L}(f_u(g_v)) + \varepsilon) \left(\log^{[q]} r + \exp^{[t]} L(r)\right)$$

$$i.e., \ \log^{[m]} T(r, f_u(g_v)) \leq (\rho^{(m,q,t)L}(f(g)) + \varepsilon) \left(\log^{[q]} r + \exp^{[t]} L(r)\right). \quad (2.3.10)$$

So from (2.3.6) and (2.3.10) it follows for all large values of r,

$$\frac{\log^{[m]} T(r, f_u(g_v))}{\log^{[l]} T(r, f_u)} \leq \frac{\rho^{(m,q,t)L}(f(g)) + \varepsilon}{\lambda^{(l,q,t)L}(f) - \varepsilon}.$$

As $\varepsilon(>0)$ is arbitrary we obtain that

$$\limsup_{r\to+\infty}\frac{\log^{[m]}T(r,f_u(g_v))}{\log^{[l]}T(r,f_u)}\leq\frac{\rho^{(m,q,t)L}(f(g))}{\lambda^{(l,q,t)L}(f)}. \tag{2.3.11}$$

Thus the theorem follows from (2.3.4), (2.3.7), (2.3.9) and (2.3.11).

Similarly, we may state the following theorem without proof for the right factor $g(z)$ of the composite function $f(g)(z)$:

Theorem 2.3.2 *Let $f(z)$ be a meromorphic function and $g(z)$ be a non constant entire function such that $0<\lambda^{(m,q,t)L}(f(g))\leq\rho^{(m,q,t)L}(f(g))<+\infty$ and $0<\lambda^{(l,q,t)L}(g)\leq\rho^{(l,q,t)L}(g)<+\infty$. Also let f_u and g_v be integer translations of $f(z)$ and $g(z)$, respectively, for $u,v\in\mathbb{N}$. Then*

$$\frac{\lambda^{(m,q,t)L}(f(g))}{\rho^{(l,q,t)L}(g)}\leq\liminf_{r\to+\infty}\frac{\log^{[m]}T(r,f_u(g_v))}{\log^{[l]}T(r,g_v)}\leq\frac{\lambda^{(m,q,t)L}(f(g))}{\lambda^{(l,q,t)L}(g)}\leq$$
$$\limsup_{r\to+\infty}\frac{\log^{[m]}T(r,f_u(g_v))}{\log^{[l]}T(r,g_v)}\leq\frac{\rho^{(m,q,t)L}(f(g))}{\lambda^{(l,q,t)L}(g)}.$$

Theorem 2.3.3 *Let $f(z)$ be a meromorphic function and $g(z)$ be a non constant entire function such that $0<\lambda^{(m,q,t)L}(f(g))\leq\rho^{(m,q,t)L}(f(g))<+\infty$ and $0<\rho^{(l,q,t)L}(f)<+\infty$. Also let f_u and g_v be integer translations of $f(z)$ and $g(z)$, respectively, for $u,v\in\mathbb{N}$. Then*

$$\liminf_{r\to+\infty}\frac{\log^{[m]}T(r,f_u(g_v))}{\log^{[l]}T(r,f_u)}\leq\frac{\rho^{(m,q,t)L}(f(g))}{\rho^{(l,q,t)L}(f)}\leq\limsup_{r\to+\infty}\frac{\log^{[m]}T(r,f_u(g_v))}{\log^{[l]}T(r,f_u)}.$$

Proof From the definition of order we get for a sequence of values of r tending to infinity,

$$\log^{[l]}T(r,f_u)\geqslant\left(\rho^{(l,q,t)L}(f_u)-\varepsilon\right)\left(\log^{[q]}r+\exp^{[t]}L(r)\right)$$
$$i.e.,\ \log^{[l]}T(r,f_u)\geqslant\left(\rho^{(l,q,t)L}(f)-\varepsilon\right)\left(\log^{[q]}r+\exp^{[t]}L(r)\right). \tag{2.3.12}$$

Now from (2.3.10) and (2.3.12) it follows for a sequence of values of r tending to infinity,

$$\frac{\log^{[m]}T(r,f_u(g_v))}{\log^{[l]}T(r,f_u)}\leq\frac{\rho^{(m,q,t)L}(f(g))+\varepsilon}{\rho^{(l,q,t)L}(f)-\varepsilon}.$$

As $\varepsilon(>0)$ is arbitrary we obtain that

$$\liminf_{r\to+\infty}\frac{\log^{[m]}T(r,f_u(g_v))}{\log^{[l]}T(r,f_u)}\leq\frac{\rho^{(m,q,t)L}(f(g))}{\rho^{(l,q,t)L}(f)}. \tag{2.3.13}$$

Again , for a sequence of values of r tending to infinity,

$$\log^{[m]}T(r,f_u(g_v))\geqslant\left(\rho^{(m,q,t)L}(f_u(g_v))-\varepsilon\right)\left(\log^{[q]}r+\exp^{[t]}L(r)\right)$$

$$i.e., \ \log^{[m]} T(r, f_u(g_v)) \geqslant (\rho^{(m,q,t)L}(f(g)) - \varepsilon)\left(\log^{[q]} r + \exp^{[t]} L(r)\right). \qquad (2.3.14)$$

So combining (2.3.3) and (2.3.14) we get for a sequence of values of r tending to infinity,

$$\frac{\log^{[m]} T(r, f_u(g_v))}{\log^{[l]} T(r, f_u)} \geqslant \frac{\rho^{(m,q,t)L}(f(g)) - \varepsilon}{\rho^{(l,q,t)L}(f) + \varepsilon}.$$

Since $\varepsilon(>0)$ is arbitrary it follows that

$$\limsup_{r\to+\infty} \frac{\log^{[m]} T(r, f_u(g_v))}{\log^{[l]} T(r, f_u)} \geqslant \frac{\rho^{(m,q,t)L}(f(g))}{\rho^{(l,q,t)L}(f)}. \qquad (2.3.15)$$

Thus the theorem follows from (2.3.13) and (2.3.15).

Theorem 2.3.4 *Let $f(z)$ be a meromorphic function and $g(z)$ be a non constant entire function such that $0 < \lambda^{(m,q,t)L}(f(g)) \leq \rho^{(m,q,t)L}(f(g)) < +\infty$ and $0 < \rho^{(l,q,t)L}(g) < +\infty$. Also let f_u and g_v be integer translations of $f(z)$ and $g(z)$, respectively, for $u, v \in \mathbb{N}$. Then*

$$\liminf_{r\to+\infty} \frac{\log^{[m]} T(r, f_u(g_v))}{\log^{[l]} T(r, g_v)} \leq \frac{\rho^{(m,q,t)L}(f(g))}{\rho^{(l,q,t)L}(g)} \leq \limsup_{r\to+\infty} \frac{\log^{[m]} T(r, f_u(g_v))}{\log^{[l]} T(r, g_v)}.$$

The proof is omitted.

The following theorem is a natural consequence of Theorem 2.3.1 and Theorem 2.3.3.

Theorem 2.3.5 *Let $f(z)$ be a meromorphic function and $g(z)$ be a non constant entire function such that $0 < \lambda^{(m,q,t)L}(f(g)) \leq \rho^{(m,q,t)L}(f(g)) < +\infty$ and $0 < \lambda^{(l,q,t)L}(f) \leq \rho^{(l,q,t)L}(f) < +\infty$. Also let f_u and g_v be integer translations of $f(z)$ and $g(z)$, respectively, for $u, v \in \mathbb{N}$. Then*

$$\liminf_{r\to+\infty} \frac{\log^{[m]} T(r, f_u(g_v))}{\log^{[l]} T(r, f_u)} \leq \min\left\{\frac{\lambda^{(m,q,t)L}(f(g))}{\lambda^{(l,q,t)L}(f)}, \frac{\rho^{(m,q,t)L}(f(g))}{\rho^{(l,q,t)L}(f)}\right\}$$
$$\leq \max\left\{\frac{\lambda^{(m,q,t)L}(f(g))}{\lambda^{(l,q,t)L}(f)}, \frac{\rho^{(m,q,t)L}(f(g))}{\rho^{(l,q,t)L}(f)}\right\} \leq \limsup_{r\to+\infty} \frac{\log^{[m]} T(r, f_u(g_v))}{\log^{[l]} T(r, f_u)}.$$

The proof is omitted.

Combining Theorem 2.3.2 and Theorem 2.3.4 we may state the following theorem.

Theorem 2.3.6 *Let $f(z)$ be a meromorphic function and $g(z)$ be a non constant entire function such that $0 < \lambda^{(m,q,t)L}(f(g)) \leq \rho^{(m,q,t)L}(f(g)) < +\infty$ and $0 < \lambda^{(l,q,t)L}(g) \leq \rho^{(l,q,t)L}(g) < +\infty$. Also let f_u and g_v be integer translations of $f(z)$ and $g(z)$, respectively, for $u, v \in \mathbb{N}$. Then*

$$\liminf_{r\to+\infty} \frac{\log^{[m]} T(r, f_u(g_v))}{\log^{[l]} T(r, g_v)} \leq \min\left\{\frac{\lambda^{(m,q,t)L}(f(g))}{\lambda^{(l,q,t)L}(g)}, \frac{\rho^{(m,q,t)L}(f(g))}{\rho^{(l,q,t)L}(g)}\right\}$$
$$\leq \max\left\{\frac{\lambda^{(m,q,t)L}(f(g))}{\lambda^{(l,q,t)L}(g)}, \frac{\rho^{(m,q,t)L}(f(g))}{\rho^{(l,q,t)L}(g)}\right\} \leq \limsup_{r\to+\infty} \frac{\log^{[m]} T(r, f_u(g_v))}{\log^{[l]} T(r, g_v)}.$$

2.4 Concluding Remark

The results of this chapter are basically inclined to prove some limiting values of $\frac{\log^{[m]} T(r,f_u(g_v))}{\log^{[l]} T(r,f_u)}$ and $\frac{\log^{[m]} T(r,f_u(g_v))}{\log^{[l]} T(r,g_v)}$. These mostly reflect the finiteness of the rations. Keeping this in mind, the theories, as well as the results of the next chapter, may have been tackled under some different conditions.

References

[1] D. Somasundaram and R.Thamizharasi, "A note on the entire functions of L-bounded index and L-type", *Indian J. Pure Appl. Math.*, Vol. 19 , No.3, pp.284-293, March 1988.

[2] T. Biswas and S. K. Datta, "Effect of integer translation on relative order and relative type of entire and meromorphic functions", *Commun. Korean Math. Soc.*, Vol. 33, No. 2, pp. 485-494, 2018.

CHAPTER 3

$(p, q, t)L$-th Order Based Some Further Results of Integer Translated Composite Entire and Meromorphic Functions

Abstract: The main purpose of this chapter is to investigate some results related to the growth rates of the composition of integer translated entire and meromorphic functions using $(p, q, t)L$-th order and $(p, q, t)L$-th lower order under certain different conditions.

Keywords: Growth, Entire function, meromorphic function, Slowly increasing function, Composition, $(p, q, t)L$-th order, $(p, q, t)L$-th lower order, integer translation.
Mathematics Subject Classification (2020): 30D30, 30D35.

3.1 Introduction

Let $\mathbb{C}$ be the set of all finite complex numbers and $f(z)$ be a meromorphic function defined on $\mathbb{C}$. Somasundaram and Thamizharasi [1] introduced the notions of L-order and L-lower order for entire functions where $L \equiv L(r)$ is a positive continuous function increasing slowly, i.e., $L(ar) \sim L(r)$ as $r \to +\infty$ for every positive constant "a". The more generalized concept of L-order and L-lower order of meromorphic functions are $(p, q, t)L$-th order and $(p, q, t)L$-th lower order, respectively. In the chapter, we establish some new results depending on the comparative growth properties of the composition of integer translated entire and meromorphic functions using $(p, q, t)L$-th order and $(p, q, t)L$-th lower order of entire and meromorphic functions under some what different conditions.

3.2 Main Results

In this section we present the main results of the chapter.

Tanmay Biswas and Chinmay Biswas

Theorem 3.2.1 *Let $f(z)$ be a meromorphic function and $g(z)$ be a non constant entire function such that $0 < \lambda^{(m,q,t)L}(f(g)) \leq \rho^{(m,q,t)L}(f(g)) < \infty$ and $0 < \lambda^{(l,q,t)L}(f) \leq \rho^{(l,q,t)L}(f) < \infty$. Also let f_u and g_v be integer translations of $f(z)$ and $g(z)$, respectively, for $u, v \in \mathbb{N}$. If $\exp^{[t]} L(r) = o\left\{\log^{[l]} T(r, f_u)\right\}$ as $r \to +\infty$ then*

$$\frac{\lambda^{(m,q,t)L}(f(g))}{\rho^{(l,q,t)L}(f)} \leq \liminf_{r\to+\infty} \frac{\log^{[m]} T(r, f_u(g_v))}{\log^{[l]} T(r, f_u) + \exp^{[t]} L(r)} \leq \frac{\lambda^{(m,q,t)L}(f(g))}{\lambda^{(l,q,t)L}(f)}$$
$$\leq \limsup_{r\to+\infty} \frac{\log^{[m]} T(r, f_u(g_v))}{\log^{[l]} T(r, f_u) + \exp^{[t]} L(r)} \leq \frac{\rho^{(m,q,t)L}(f(g))}{\lambda^{(l,q,t)L}(f)}.$$

Proof From Definition 1.2.22 and in view of Lemma 2.2.2, we have for all sufficiently large positive numbers of r that

$$\log^{[m]} T(r, f_u(g_v)) \geq \left(\lambda^{(m,q,t)L}(f_u(g_v)) - \varepsilon\right)\left(\log^{[q]} r + \exp^{[t]} L(r)\right)$$
$$i.e., \ \log^{[m]} T(r, f_u(g_v)) \geq \left(\lambda^{(m,q,t)L}(f(g)) - \varepsilon\right)\left(\log^{[q]} r + \exp^{[t]} L(r)\right), \quad (3.2.1)$$

$$\log^{[l]} T(r, f_u) \geq \left(\lambda^{(l,q,t)L}(f_u) - \varepsilon\right)\left(\log^{[q]} r + \exp^{[t]} L(r)\right)$$
$$i.e., \ \log^{[l]} T(r, f_u) \geq \left(\lambda^{(l,q,t)L}(f) - \varepsilon\right)\left(\log^{[q]} r + \exp^{[t]} L(r)\right), \quad (3.2.2)$$

$$\log^{[m]} T(r, f_u(g_v)) \leq \left(\rho^{(m,q,t)L}(f_u(g_v)) + \varepsilon\right)\left(\log^{[q]} r + \exp^{[t]} L(r)\right)$$
$$i.e., \ \log^{[m]} T(r, f_u(g_v)) \leq \left(\rho^{(m,q,t)L}(f(g)) + \varepsilon\right)\left(\log^{[q]} r + \exp^{[t]} L(r)\right) \quad (3.2.3)$$

and

$$\log^{[l]} T(r, f_u) \leq \left(\rho^{(l,q,t)L}(f_u) + \varepsilon\right)\left(\log^{[q]} r + \exp^{[t]} L(r)\right)$$
$$i.e., \ \log^{[l]} T(r, f_u) \leq \left(\rho^{(l,q,t)L}(f) + \varepsilon\right)\left(\log^{[q]} r + \exp^{[t]} L(r)\right). \quad (3.2.4)$$

Also for a sequence of positive numbers of r tending to infinity

$$\log^{[m]} T(r, f_u(g_v)) \leq \left(\lambda^{(m,q,t)L}(f_u(g_v)) + \varepsilon\right)\left(\log^{[q]} r + \exp^{[t]} L(r)\right)$$
$$i.e., \ \log^{[m]} T(r, f_u(g_v)) \leq \left(\lambda^{(m,q,t)L}(f(g)) + \varepsilon\right)\left(\log^{[q]} r + \exp^{[t]} L(r)\right), \quad (3.2.5)$$

$$\log^{[l]} T(r, f_u) \leq \left(\lambda^{(l,q,t)L}(f_u) + \varepsilon\right)\left(\log^{[q]} r + \exp^{[t]} L(r)\right)$$
$$i.e., \ \log^{[l]} T(r, f_u) \leq \left(\lambda^{(l,q,t)L}(f) + \varepsilon\right)\left(\log^{[q]} r + \exp^{[t]} L(r)\right), \quad (3.2.6)$$

$$\log^{[m]} T(r, f_u(g_v)) \geq \left(\rho^{(m,q,t)L}(f_u(g_v)) - \varepsilon\right)\left(\log^{[q]} r + \exp^{[t]} L(r)\right)$$

$$i.e., \ \log^{[m]} T(r, f_u(g_v)) \geq \left(\rho^{(m,q,t)L}(f(g)) - \varepsilon\right)\left(\log^{[q]} r + \exp^{[t]} L(r)\right) \tag{3.2.7}$$

and

$$\log^{[l]} T(r, f_u) \geq \left(\rho^{(l,q,t)L}(f_u) - \varepsilon\right)\left(\log^{[q]} r + \exp^{[t]} L(r)\right)$$

$$i.e., \ \log^{[l]} T(r, f_u) \geq \left(\rho^{(l,q,t)L}(f) - \varepsilon\right)\left(\log^{[q]} r + \exp^{[t]} L(r)\right). \tag{3.2.8}$$

Now from (3.2.4) we get for all sufficiently large positive numbers of r that

$$\frac{\log^{[l]} T(r, f_u)}{\left(\rho^{(l,q,t)L}(f) + \varepsilon\right)} \leq \log^{[q]} r + \exp^{[t]} L(r). \tag{3.2.9}$$

Now from (3.2.1) and (3.2.9), it follows for all sufficiently large positive numbers of r that

$$\log^{[m]} T(r, f_u(g_v)) \geq \frac{\left(\lambda^{(m,q,t)L}(f(g)) - \varepsilon\right)}{\left(\rho^{(l,q,t)L}(f) + \varepsilon\right)} \log^{[l]} T(r, f_u)$$

$$i.e., \ \frac{\log^{[m]} T(r, f_u(g_v))}{\log^{[l]} T(r, f_u) + \exp^{[t]} L(r)} \geq \frac{\left(\lambda^{(m,q,t)L}(f(g)) - \varepsilon\right)}{\left(\rho^{(l,q,t)L}(f) + \varepsilon\right)} \cdot \frac{\log^{[l]} T(r, f_u)}{\log^{[l]} T(r, f_u) + \exp^{[t]} L(r)}$$

$$i.e., \ \frac{\log^{[m]} T(r, f_u(g_v))}{\log^{[l]} T(r, f_u) + \exp^{[t]} L(r)} \geq \frac{\frac{\left(\lambda^{(m,q,t)L}(f(g)) - \varepsilon\right)}{\left(\rho^{(l,q,t)L}(f) + \varepsilon\right)}}{1 + \frac{\exp^{[t]} L(r)}{\log^{[l]} T(r, f_u)}}.$$

Since $\exp^{[t]} L(r) = o\left\{\log^{[l]} T(r, f_u)\right\}$ as $r \to +\infty$, it follows from above that

$$\liminf_{r \to +\infty} \frac{\log^{[m]} T(r, f_u(g_v))}{\log^{[l]} T(r, f_u) + \exp^{[t]} L(r)} \geq \frac{\left(\lambda^{(m,q,t)L}(f(g)) - \varepsilon\right)}{\left(\rho^{(l,q,t)L}(f) + \varepsilon\right)}. \tag{3.2.10}$$

As $\varepsilon\,(> 0)$ is arbitrary, we get from (3.2.10) that

$$\liminf_{r \to +\infty} \frac{\log^{[m]} T(r, f_u(g_v))}{\log^{[l]} T(r, f_u) + \exp^{[t]} L(r)} \geq \frac{\lambda^{(m,q,t)L}(f(g))}{\rho^{(l,q,t)L}(f)}. \tag{3.2.11}$$

Again from (3.2.2), we obtain for all sufficiently large positive numbers of r that

$$\frac{\log^{[l]} T(r, f_u)}{\left(\lambda^{(l,q,t)L}(f) - \varepsilon\right)} \geq \log^{[q]} r + \exp^{[t]} L(r). \tag{3.2.12}$$

From (3.2.5) and (3.2.12), it follows for a sequence of positive numbers of r tending to infinity that

$$\log^{[m]} T(r, f_u(g_v)) \leq \frac{\left(\lambda^{(m,q,t)L}(f(g)) + \varepsilon\right)}{\left(\lambda^{(l,q,t)L}(f) - \varepsilon\right)} \log^{[l]} T(r, f_u)$$

$$i.e., \frac{\log^{[m]} T(r, f_u(g_v))}{\log^{[l]} T(r, f_u) + \exp^{[t]} L(r)} \leq \frac{\left(\lambda^{(m,q,t)L}(f(g)) + \varepsilon\right)}{\left(\lambda^{(l,q,t)L}(f) - \varepsilon\right)} \cdot \frac{\log^{[l]} T(r, f_u)}{\log^{[l]} T(r, f_u) + \exp^{[t]} L(r)}$$

$$i.e., \frac{\log^{[m]} T(r, f_u(g_v))}{\log^{[l]} T(r, f_u) + \exp^{[t]} L(r)} \leq \frac{\frac{\left(\lambda^{(m,q,t)L}(f(g)) + \varepsilon\right)}{\left(\lambda^{(l,q,t)L}(f) - \varepsilon\right)}}{1 + \frac{\exp^{[t]} L(r)}{\log^{[l]} T(r, f_u)}}.$$

As $\exp^{[t]} L(r) = o\left\{\log^{[l]} T(r, f_u)\right\}$ as $r \to +\infty$ we get from above that

$$\liminf_{r \to +\infty} \frac{\log^{[m]} T(r, f_u(g_v))}{\log^{[l]} T(r, f_u) + \exp^{[t]} L(r)} \leq \frac{\left(\lambda^{(m,q,t)L}(f(g)) + \varepsilon\right)}{\left(\lambda^{(l,q,t)L}(f) - \varepsilon\right)}. \tag{3.2.13}$$

Since $\varepsilon (> 0)$ is arbitrary, it follows from (3.2.13) that

$$\liminf_{r \to +\infty} \frac{\log^{[m]} T(r, f_u(g_v))}{\log^{[l]} T(r, f_u) + \exp^{[t]} L(r)} \leq \frac{\lambda^{(m,q,t)L}(f(g))}{\lambda^{(l,q,t)L}(f)}. \tag{3.2.14}$$

Also from (3.2.6), we obtain for a sequence of positive numbers of r tending to infinity that

$$\frac{\log^{[l]} T(r, f_u)}{\left(\lambda^{(l,q,t)L}(f) + \varepsilon\right)} \leq \log^{[q]} r + \exp^{[t]} L(r). \tag{3.2.15}$$

Now from (3.2.1) and (3.2.15), we get for a sequence of positive numbers of r tending to infinity that

$$\log^{[m]} T(r, f_u(g_v)) \geq \frac{\left(\lambda^{(m,q,t)L}(f(g)) - \varepsilon\right)}{\left(\lambda^{(l,q,t)L}(f) + \varepsilon\right)} \log^{[l]} T(r, f_u)$$

$$i.e., \frac{\log^{[m]} T(r, f_u(g_v))}{\log^{[l]} T(r, f_u) + \exp^{[t]} L(r)} \geq \frac{\left(\lambda^{(m,q,t)L}(f(g)) - \varepsilon\right)}{\left(\lambda^{(l,q,t)L}(f) + \varepsilon\right)} \cdot \frac{\log^{[l]} T(r, f_u)}{\log^{[l]} T(r, f_u) + \exp^{[t]} L(r)}$$

$$i.e., \frac{\log^{[m]} T(r, f_u(g_v))}{\log^{[l]} T(r, f_u) + \exp^{[t]} L(r)} \geq \frac{\frac{\left(\lambda^{(m,q,t)L}(f(g)) - \varepsilon\right)}{\left(\lambda^{(l,q,t)L}(f) + \varepsilon\right)}}{1 + \frac{\exp^{[t]} L(r)}{\log^{[l]} T(r, f_u)}}.$$

In view of the condition $\exp^{[t]} L(r) = o\left\{\log^{[l]} T(r, f_u)\right\}$ as $r \to +\infty$ we obtain from above that

$$\limsup_{r \to +\infty} \frac{\log^{[m]} T(r, f_u(g_v))}{\log^{[l]} T(r, f_u) + \exp^{[t]} L(r)} \geq \frac{\lambda^{(m,q,t)L}(f(g)) - \varepsilon}{\left(\lambda^{(l,q,t)L}(f) + \varepsilon\right)}. \tag{3.2.16}$$

Since $\varepsilon (> 0)$ is arbitrary, it follows from (3.2.16) that

$$\limsup_{r \to +\infty} \frac{\log^{[m]} T(r, f_u(g_v))}{\log^{[l]} T(r, f_u) + \exp^{[t]} L(r)} \geq \frac{\lambda^{(m,q,t)L}(f(g))}{\lambda^{(l,q,t)L}(f)}. \tag{3.2.17}$$

Again from (3.2.2), we get for all sufficiently large positive numbers of r that

$$\frac{\log^{[l]} T(r, f_u)}{A\left(\lambda^{(l,q,t)L}(f) - \varepsilon\right)} \geq \log^{[q]} r + \exp^{[t]} L(r). \tag{3.2.18}$$

Combining (3.2.3) and (3.2.18), it follows for all sufficiently large positive numbers of r that

$$\log^{[m]} T(r, f_u(g_v)) \leq \frac{\left(\rho^{(m,q,t)L}(f(g)) + \varepsilon\right)}{\left(\lambda^{(l,q,t)L}(f) - \varepsilon\right)} \log^{[l]} T(r, f_u)$$

$$i.e., \ \frac{\log^{[m]} T(r, f_u(g_v))}{\log^{[l]} T(r, f_u) + \exp^{[t]} L(r)} \leq \frac{\rho^{(m,q,t)L}(f(g)) + \varepsilon}{\left(\lambda^{(l,q,t)L}(f) - \varepsilon\right)} \cdot \frac{\log^{[l]} T(r, f_u)}{\log^{[l]} T(r, f_u) + \exp^{[t]} L(r)}$$

$$i.e., \ \frac{\log^{[m]} T(r, f_u(g_v))}{\log^{[l]} T(r, f_u) + \exp^{[t]} L(r)} \leq \frac{\frac{\rho^{(m,q,t)L}(f(g)) + \varepsilon}{\left(\lambda^{(l,q,t)L}(f) - \varepsilon\right)}}{1 + \frac{\exp^{[t]} L(r)}{\log^{[l]} T(r, f_u)}}.$$

Using $\exp^{[t]} L(r) = o\left\{\log^{[l]} T(r, f_u)\right\}$ as $r \to +\infty$ we obtain from above that

$$\limsup_{r \to +\infty} \frac{\log^{[m]} T(r, f_u(g_v))}{\log^{[l]} T(r, f_u) + \exp^{[t]} L(r)} \leq \frac{\rho^{(m,q,t)L}(f(g)) + \varepsilon}{\left(\lambda^{(l,q,t)L}(f) - \varepsilon\right)}. \tag{3.2.19}$$

As $\varepsilon (> 0)$ is arbitrary, it follows from (3.2.19) that

$$\limsup_{r \to +\infty} \frac{\log^{[m]} T(r, f_u(g_v))}{\log^{[l]} T(r, f_u) + \exp^{[t]} L(r)} \leq \frac{\rho^{(m,q,t)L}(f(g))}{\lambda^{(l,q,t)L}(f)}. \tag{3.2.20}$$

Thus the theorem follows from (3.2.11), (3.2.14), (3.2.17) and (3.2.20).

In view of Theorem 3.2.1 the following theorem may be carried out:

Theorem 3.2.2 *Let $f(z)$ be a meromorphic function and $g(z)$ be a non constant entire function such that $0 < \lambda^{(m,q,t)L}(f(g)) \leq \rho^{(m,q,t)L}(f(g)) < \infty$, $0 < \lambda^{(l,q,t)L}(g) \leq \rho^{(l,q,t)L}(g) < \infty$. Also let f_u and g_v be integer translations of $f(z)$ and $g(z)$, respectively, for $u, v \in \mathbb{N}$. If $\exp^{[t]} L(r) = o\left\{\log^{[l]} T(r, g_v)\right\}$ as $r \to +\infty$ then*

$$\frac{\lambda^{(m,q,t)L}(f(g))}{\rho^{(l,q,t)L}(g)} \leq \liminf_{r \to +\infty} \frac{\log^{[m]} T(r, f_u(g_v))}{\log^{[l]} T(r, g_v) + \exp^{[t]} L(r)} \leq \frac{\lambda^{(m,q,t)L}(f(g))}{\lambda^{(l,q,t)L}(g)}$$
$$\leq \limsup_{r \to +\infty} \frac{\log^{[m]} T(r, f_u(g_v))}{\log^{[l]} T(r, g_v) + \exp^{[t]} L(r)} \leq \frac{\rho^{(m,q,t)L}(f(g))}{\lambda^{(l,q,t)L}(g)}.$$

Theorem 3.2.3 *Let $f(z)$ be a meromorphic function and $g(z)$ be a non constant entire function such that $0 < \lambda^{(m,q,t)L}(f(g)) \leq \rho^{(m,q,t)L}(f(g)) < \infty$ and $0 < \rho^{(l,q,t)L}(f) < \infty$. Also let f_u and g_v be integer translations of $f(z)$ and $g(z)$, respectively, for $u, v \in \mathbb{N}$. If $\exp^{[t]} L(r) = o\left\{\log^{[l]} T(r, f_u)\right\}$ as $r \to +\infty$ then*

$$\liminf_{r \to +\infty} \frac{\log^{[m]} T(r, f_u(g_v))}{\log^{[l]} T(r, f_u) + \exp^{[t]} L(r)} \leq \frac{\rho^{(m,q,t)L}(f(g))}{\rho^{(l,q,t)L}(f)} \leq \limsup_{r \to +\infty} \frac{\log^{[m]} T(r, f_u(g_v))}{\log^{[l]} T(r, f_u) + \exp^{[t]} L(r)}.$$

Proof From Definition 1.2.22 and in view of Lemma 2.2.2, we get for a sequence of positive numbers of r tending to infinity that

$$\log^{[l]} T(r, f_u) \geq \left(\rho^{(l,q,t)L}(f_u) - \varepsilon\right) \left(\log^{[q]} r + \exp^{[t]} L(r)\right)$$

$$i.e., \ \log^{[l]} T(r, f_u) \geq \left(\rho^{(l,q,t)L}(f) - \varepsilon\right) \left(\log^{[q]} r + \exp^{[t]} L(r)\right)$$

$$i.e., \ \frac{\log^{[l]} T(r, f_u)}{\left(\rho^{(l,q,t)L}(f) - \varepsilon\right)} \geq \left(\log^{[q]} r + \exp^{[t]} L(r)\right). \tag{3.2.21}$$

Now from (3.2.3) and (3.2.21), it follows for a sequence of positive numbers of r tending to infinity that

$$\log^{[m]} T(r, f_u(g_v)) \leq \frac{\left(\rho^{(m,q,t)L}(f(g)) + \varepsilon\right)}{\left(\rho^{(l,q,t)L}(f) - \varepsilon\right)} \log^{[l]} T(r, f_u)$$

$$i.e., \ \frac{\log^{[m]} T(r, f_u(g_v))}{\log^{[l]} T(r, f_u) + \exp^{[t]} L(r)} \leq \frac{\rho^{(m,q,t)L}(f(g)) + \varepsilon}{\left(\rho^{(l,q,t)L}(f) - \varepsilon\right)} \cdot \frac{\log^{[l]} T(r, f_u)}{\log^{[l]} T(r, f_u) + \exp^{[t]} L(r)}$$

$$i.e., \ \frac{\log^{[m]} T(r, f_u(g_v))}{\log^{[l]} T(r, f_u) + \exp^{[t]} L(r)} \leq \frac{\frac{\rho^{(m,q,t)L}(f(g)) + \varepsilon}{\left(\rho^{(l,q,t)L}(f) - \varepsilon\right)}}{1 + \frac{\exp^{[t]} L(r)}{\log^{[l]} T(r, f_u)}}.$$

Using $\exp^{[t]} L(r) = o\left\{\log^{[l]} T(r, f_u)\right\}$ as $r \to +\infty$ we obtain from above that

$$\liminf_{r \to +\infty} \frac{\log^{[m]} T(r, f_u(g_v))}{\log^{[l]} T(r, f_u) + \exp^{[t]} L(r)} \leq \frac{\rho^{(m,q,t)L}(f(g)) + \varepsilon}{\left(\rho^{(l,q,t)L}(f) - \varepsilon\right)}. \tag{3.2.22}$$

As $\varepsilon (> 0)$ is arbitrary, it follows from (3.2.22) that

$$\liminf_{r \to +\infty} \frac{\log^{[m]} T(r, f_u(g_v))}{\log^{[l]} T(r, f_u) + \exp^{[t]} L(r)} \leq \frac{\rho^{(m,q,t)L}(f(g))}{\rho^{(l,q,t)L}(f)}. \tag{3.2.23}$$

Again for a sequence of positive numbers of r tending to infinity,

$$\log^{[m]} T(r, f_u(g_v)) \geq \left(\rho^{(m,q,t)L}(f_u(g_v)) - \varepsilon\right) \left(\log^{[q]} r + \exp^{[t]} L(r)\right)$$

$$i.e., \ \log^{[m]} T(r, f_u(g_v)) \geq \left(\rho^{(m,q,t)L}(f(g)) - \varepsilon\right) \left(\log^{[q]} r + \exp^{[t]} L(r)\right). \tag{3.2.24}$$

So combining (3.2.9) and (3.2.24), we get for a sequence of positive numbers of r tending to infinity that

$$\log^{[m]} T(r, f_u(g_v)) \geq \frac{\left(\rho^{(m,q,t)L}(f(g)) - \varepsilon\right)}{\left(\rho^{(l,q,t)L}(f) + \varepsilon\right)} \log^{[l]} T(r, f_u)$$

$$i.e., \ \frac{\log^{[m]} T(r, f_u(g_v))}{\log^{[l]} T(r, f_u) + \exp^{[t]} L(r)} \geq \frac{\rho^{(m,q,t)L}(f(g)) - \varepsilon}{\left(\rho^{(l,q,t)L}(f) + \varepsilon\right)} \cdot \frac{\log^{[l]} T(r, f_u)}{\log^{[l]} T(r, f_u) + \exp^{[t]} L(r)}$$

$$i.e., \frac{\log^{[m]} T\left(r, f_u(g_v)\right)}{\log^{[l]} T\left(r, f_u\right) + \exp^{[t]} L\left(r\right)} \geq \frac{\frac{\rho^{(m,q,t)L}(f(g))-\varepsilon}{\left(\rho^{(l,q,t)L}(f)+\varepsilon\right)}}{1 + \frac{\exp^{[t]} L(r)}{\log^{[l]} T(r,f_u)}}.$$

Since $\exp^{[t]} L\left(r\right) = o\left\{\log^{[l]} T\left(r, f_u\right)\right\}$ as $r \to +\infty$, it follows from above that

$$\limsup_{r\to+\infty} \frac{\log^{[m]} T\left(r, f_u(g_v)\right)}{\log^{[l]} T\left(r, f_u\right) + \exp^{[t]} L\left(r\right)} \geq \frac{\left(\rho^{(m,q,t)L}\left(f(g)\right) - \varepsilon\right)}{\left(\rho^{(l,q,t)L}\left(f\right) + \varepsilon\right)}. \tag{3.2.25}$$

As $\varepsilon\,(> 0)$ is arbitrary, we get from (3.2.25) that

$$\limsup_{r\to+\infty} \frac{\log^{[m]} T\left(r, f_u(g_v)\right)}{\log^{[l]} T\left(r, f_u\right) + \exp^{[t]} L\left(r\right)} \geq \frac{\rho^{(m,q,t)L}\left(f(g)\right)}{\rho^{(l,q,t)L}\left(f\right)}. \tag{3.2.26}$$

Thus the theorem follows from (3.2.23) and (3.2.26).

Theorem 3.2.4 *Let $f\left(z\right)$ be a meromorphic function and $g\left(z\right)$ be a non constant entire function such that $0 < \lambda^{(m,q,t)L}\left(f(g)\right) \leq \rho^{(m,q,t)L}\left(f(g)\right) < \infty$, $0 < \rho^{(l,q,t)L}\left(g\right) < \infty$. Also let f_u and g_v be integer translations of $f\left(z\right)$ and $g\left(z\right)$, respectively, for $u, v \in \mathbb{N}$. If $\exp^{[t]} L\left(r\right) = o\left\{\log^{[l]} T\left(r, g_v\right)\right\}$ as $r \to +\infty$ then*

$$\liminf_{r\to+\infty} \frac{\log^{[m]} T\left(r, f_u(g_v)\right)}{\log^{[l]} T\left(r, g_v\right) + \exp^{[t]} L\left(r\right)} \leq \frac{\rho^{(m,q,t)L}\left(f(g)\right)}{\rho^{(l,q,t)L}\left(g\right)} \leq \limsup_{r\to+\infty} \frac{\log^{[m]} T\left(r, f_u(g_v)\right)}{\log^{[l]} T\left(r, g_v\right) + \exp^{[t]} L\left(r\right)}.$$

The proof is omitted as it can be carried out in the line of Theorem 3.2.3.

The following theorem is a natural consequence of Theorem 3.2.1 and Theorem 3.2.3:

Theorem 3.2.5 *Let $f\left(z\right)$ be a meromorphic function and $g\left(z\right)$ be a non constant entire function such that $0 < \lambda^{(m,q,t)L}\left(f(g)\right) \leq \rho^{(m,q,t)L}\left(f(g)\right) < \infty$ and $0 < \lambda^{(l,q,t)L}\left(f\right) \leq \rho^{(l,q,t)L}\left(f\right) < \infty$. Also let f_u and g_v be integer translations of $f\left(z\right)$ and $g\left(z\right)$, respectively, for $u, v \in \mathbb{N}$. If $\exp^{[t]} L\left(r\right) = o\left\{\log^{[l]} T\left(r, f_u\right)\right\}$ as $r \to +\infty$ then*

$$\liminf_{r\to+\infty} \frac{\log^{[m]} T\left(r, f_u(g_v)\right)}{\log^{[l]} T\left(r, f_u\right) + \exp^{[t]} L\left(r\right)} \leq \min\left\{\frac{\lambda^{(m,q,t)L}\left(f(g)\right)}{\lambda^{(l,q,t)L}\left(f\right)}, \frac{\rho^{(m,q,t)L}\left(f(g)\right)}{\rho^{(l,q,t)L}\left(f\right)}\right\}$$
$$\leq \max\left\{\frac{\lambda^{(m,q,t)L}\left(f(g)\right)}{\lambda^{(l,q,t)L}\left(f\right)}, \frac{\rho^{(m,q,t)L}\left(f(g)\right)}{\rho^{(l,q,t)L}\left(f\right)}\right\} \leq \limsup_{r\to+\infty} \frac{\log^{[m]} T\left(r, f_u(g_v)\right)}{\log^{[l]} T\left(r, f_u\right) + \exp^{[t]} L\left(r\right)}.$$

The proof is omitted.

Combining Theorem 3.2.2 and Theorem 3.2.4, we may state the following theorem:

Theorem 3.2.6 *Let $f\left(z\right)$ be a meromorphic function and $g\left(z\right)$ be a non constant entire function such that $0 < \lambda^{(m,q,t)L}\left(f(g)\right) \leq \rho^{(m,q,t)L}\left(f(g)\right) < \infty$, $0 < \lambda^{(l,q,t)L}\left(g\right) \leq \rho^{(l,q,t)L}\left(g\right)$*

$< \infty$. *Also let* f_u *and* g_v *be integer translations of* $f(z)$ *and* $g(z)$, *respectively, for* $u, v \in \mathbb{N}$. *If* $\exp^{[t]} L(r) = o\left\{\log^{[l]} T(r, g_v)\right\}$ *as* $r \to +\infty$ *then*

$$\liminf_{r\to+\infty} \frac{\log^{[m]} T(r, f_u(g_v))}{\log^{[l]} T(r, g_v) + \exp^{[t]} L(r)} \leq \min\left\{\frac{\lambda^{(m,q,t)L}(f(g))}{\lambda^{(l,q,t)L}(g)}, \frac{\rho^{(m,q,t)L}(f(g))}{\rho^{(l,q,t)L}(g)}\right\}$$

$$\leq \max\left\{\frac{\lambda^{(m,q,t)L}(f(g))}{\lambda^{(l,q,t)L}(g)}, \frac{\rho^{(m,q,t)L}(f(g))}{\rho^{(l,q,t)L}(g)}\right\} \leq \limsup_{r\to+\infty} \frac{\log^{[m]} T(r, f_u(g_v))}{\log^{[l]} T(r, g_v) + \exp^{[t]} L(r)}.$$

3.3 Concluding Remark

The results of this chapter are mainly based on the investigations of some basic growth properties of composite integer translated entire and meromorphic functions on the basis of $(p, q, t)L$-th order and $(p, q, t)L$-th lower order. As a further area of research in the area of growth properties of composite entire and meromorphic functions, $(p, q, t)L$-th type and $(p, q, t)L$-th weak type play an important role, and as a consequence, one may see some more powerful results in the next chapter.

References

[1] D. Somasundaram and R.Thamizharasi, "A note on the entire functions of L-bounded index and L-type", *Indian J. Pure Appl. Math.*, Vol. 19 , No.3 (March 1988), pp.284-293.

CHAPTER 4

$(p, q, t)L$-th Type and $(p, q, t)L$-th Weak Type Based Some Growth Properties of Composite Entire and Meromorphic Functions on the Basis of Their Integer Translation

Abstract: The main objective of this chapter is to investigate some results related to the growth rates of the composition of integer translated entire and meromorphic functions using $(p, q, t)L$-th type and $(p, q, t)L$-th weak type.

Keywords: Entire function, meromorphic function, $(p, q, t)L$-th type, $(p, q, t)L$-th weak type, integer translation.
Mathematics Subject Classification (2020): 30D30, 30D35.

4.1 Introduction

Let $\mathbb{C}$ be the set of all finite complex numbers and $f(z)$ be a meromorphic function defined on $\mathbb{C}$. Somasundaram and Thamizharasi [1] introduced the notions of L-order and L-lower order for entire functions where $L \equiv L(r)$ is a positive continuous function increasing slowly, *i.e.*, $L(ar) \sim L(r)$ as $r \to +\infty$ for every positive constant "a". The more generalized concept of L-order and L-lower order of meromorphic functions are $(p, q, t)L$-th order and $(p, q, t)L$-th lower order, respectively. In order to compare the growth of entire or meromorphic functions having the same $(p, q, t)L$-th order or $(p, q, t)L$-th lower order, one may give the definitions of $(p, q, t)L$-th type and $(p, q, t)L$-th weak type of entire or meromorphic functions. In this chapter, we establish some new results depending on the comparative growth properties of composition of the integer translated entire and meromorphic functions using $(p, q, t)L$-th type and $(p, q, t)L$-th weak type of entire and meromorphic functions.

Tanmay Biswas and Chinmay Biswas

4.2 Lemmas

In this section, we present some lemmas which will be needed in the sequel.

Lemma 4.2.1 *[2] Let $f(z)$ be a meromorphic function. If $f_n(z) = f(z+n)$ for $n \in \mathbb{N}$ then*

$$\lim_{r\to+\infty} \frac{T(r, f_n)}{T(r, f)} = n.$$

Lemma 4.2.2 *Let $f(z)$ be a meromorphic function. If $f_n(z) = f(z+n)$ for $n \in \mathbb{N}$ then*

$$(i)\ \sigma^{(p,q,t)L}(f_n) = \begin{cases} n \cdot \sigma^{(p,q,t)L}(f) & \text{for } p = 1 \\ \sigma^{(p,q,t)L}(f) & \text{for } p > 1 \end{cases}$$

and

$$(ii)\ \overline{\sigma}^{(p,q,t)L}(f_n) = \begin{cases} n \cdot \overline{\sigma}^{(p,q,t)L}(f) & \text{for } p = 1 \\ \overline{\sigma}^{(p,q,t)L}(f) & \text{for } p > 1. \end{cases}$$

Proof By Lemma 4.2.1 and Lemma 2.2.2, we get that

$$\begin{aligned} \sigma^{(1,q,t)L}(f_n) &= \limsup_{r\to+\infty} \frac{T(r, f_n)}{\left[\log^{[q-1]} r \cdot \exp^{[t+1]} L(r)\right]^{\rho^{(1,q,t)L}(f_n)}} \\ &= \lim_{r\to+\infty} \frac{T(r, f_n)}{T(r, f)} \cdot \limsup_{r\to+\infty} \frac{T(r, f)}{\left[\log^{[q-1]} r \cdot \exp^{[t+1]} L(r)\right]^{\rho^{(1,q,t)L}(f)}} \\ &= n \cdot \sigma^{(1,q,t)L}(f). \end{aligned}$$

Also for $p > 1$, in view of Lemma 4.2.1, $\lim_{r\to+\infty} \frac{\log^{[p-1]} T(r,f_n)}{\log^{[p-1]} T(r,f)}$ exists and is equal to 1. Therefore in view of Lemma 2.2.2 we obtain that

$$\begin{aligned} \sigma^{(p,q,t)L}(f_n) &= \limsup_{r\to+\infty} \frac{\log^{[p-1]} T(r, f_n)}{\left[\log^{[q-1]} r \cdot \exp^{[t+1]} L(r)\right]^{\rho^{(p,q,t)L}(f_n)}} \\ &= \lim_{r\to+\infty} \frac{\log^{[p-1]} T(r, f_n)}{\log^{[p-1]} T(r, f)} \cdot \limsup_{r\to+\infty} \frac{\log^{[p-1]} T(r, f)}{\left[\log^{[q-1]} r \cdot \exp^{[t+1]} L(r)\right]^{\rho^{(p,q,t)L}(f)}} \\ &= \sigma^{(p,q,t)L}(f). \end{aligned}$$

In a similar manner,

$$\overline{\sigma}^{(p,q,t)L}(f_n) = n \cdot \overline{\sigma}^{(p,q,t)L}(f) \text{ for } p = 1$$

$$\text{and } \overline{\sigma}^{(p,q,t)L}(f_n) = \overline{\sigma}^{(p,q,t)L}(f) \text{ otherwise.}$$

Thus the lemma follows.

Lemma 4.2.3 *Let $f(z)$ be a meromorphic function. If $f_n(z) = f(z+n)$ for $n \in \mathbb{N}$ then*

$$(i)\ \tau^{(p,q,t)L}(f_n) = \begin{cases} n \cdot \tau^{(p,q,t)L}(f) & \text{for } p = 1 \\ \tau^{(p,q,t)L}(f) & \text{for } p > 1 \end{cases}$$

and

$$(ii)\ \overline{\tau}^{(p,q,t)L}(f_n) = \begin{cases} n \cdot \overline{\tau}^{(p,q,t)L}(f) & \text{for } p = 1 \\ \overline{\tau}^{(p,q,t)L}(f) & \text{for } p > 1. \end{cases}$$

The proof of Lemma 4.2.3 is omitted as it can easily be carried out in the line of Lemma 4.2.2.

4.3 Main Results

In this section, we present the main results of the chapter.

Theorem 4.3.1 *Let $f(z)$ be a meromorphic function and $g(z)$ be a non constant entire function such that $0 < \overline{\sigma}^{(m,q,t)L}(f(g)) \le \sigma^{(m,q,t)L}(f(g)) < \infty$, $0 < \overline{\sigma}^{(l,q,t)L}(f) \le \sigma^{(l,q,t)L}(f) < \infty$, $\rho^{(m,q,t)L}(f(g)) = \rho^{(l,q,t)L}(f)$. Also let f_u and g_v be integer translations of $f(z)$ and $g(z)$, respectively, for $u, v \in \mathbb{N}$. If $f_u(g_v) = h_t$, where h is a meromorphic function and $t \in \mathbb{N}$, then*

$$\frac{t \cdot \overline{\sigma}^{(1,q,t)L}(f(g))}{u \cdot \sigma^{(1,q,t)L}(f)} \le \liminf_{r\to+\infty} \frac{T(r, f_u(g_v))}{T(r, f_u)} \le \frac{t \cdot \overline{\sigma}^{(1,q,t)L}(f(g))}{u \cdot \overline{\sigma}^{(1,q,t)L}(f)} \le \limsup_{r\to+\infty} \frac{T(r, f_u(g_v))}{T(r, f_u)} \le \frac{t \cdot \sigma^{(1,q,t)L}(f(g))}{u \cdot \overline{\sigma}^{(1,q,t)L}(f)}$$

and

$$\frac{\overline{\sigma}^{(m,q,t)L}(f(g))}{\sigma^{(l,q,t)L}(f)} \le \liminf_{r\to+\infty} \frac{\log^{[m-1]} T(r, f_u(g_v))}{\log^{[l-1]} T(r, f_u)} \le \frac{\overline{\sigma}^{(m,q,t)L}(f(g))}{\overline{\sigma}^{(l,q,t)L}(f)} \le \limsup_{r\to+\infty} \frac{\log^{[m-1]} T(r, f_u(g_v))}{\log^{[l-1]} T(r, f_u)} \le \frac{\sigma^{(m,q,t)L}(f(g))}{\overline{\sigma}^{(l,q,t)L}(f)}$$

for $m > 1$ and $l > 1$.

Proof By the procedure of establishing (2.1.4) we can express

$$T(r, f_u(g_v)) = tT(r, f(g)) + \sum_t (e_t + e_t'),$$

where $e_t, e_t' \to 0$ as $r \to +\infty$. Therefore

$$\lim_{r\to+\infty} \frac{T(r, f_u(g_v))}{T(r, f(g))} = t.$$

So in view of Lemma 4.2.2, we get that

$$\sigma^{(1,q,t)L}\left(f_u(g_v)\right) = t \cdot \sigma^{(1,q,t)L}\left(f(g)\right) \text{ and}$$
$$\overline{\sigma}^{(1,q,t)L}\left(f_u(g_v)\right) = t \cdot \overline{\sigma}^{(1,q,t)L}\left(f(g)\right). \tag{4.3.1}$$

Similarly, when $m > 1$, in view of Lemma 4.2.2, we obtain that

$$\sigma^{(m,q,t)L}\left(f_u(g_v)\right) = \sigma^{(m,q,t)L}\left(f(g)\right) \text{ and } \overline{\sigma}^{(m,q,t)L}\left(f_u(g_v)\right) = \overline{\sigma}^{(m,q,t)L}\left(f(g)\right) \tag{4.3.2}$$

Therefore in view of Lemma 2.2.2, Lemma 4.2.2, (4.3.1) and (2.3.1) we have for arbitrary positive ε and for all large values of r that

$$T(r, f_u(g_v)) \geqslant \left(\overline{\sigma}^{(1,q,t)L}\left(f_u(g_v)\right) - \varepsilon\right)\left(\log^{[q-1]} r \cdot \exp^{[t+1]} L(r)\right)^{\rho^{(1,q,t)L}(f_u(g_v))}$$

$$i.e.,\ T(r, f_u(g_v)) \geqslant \left(t \cdot \overline{\sigma}^{(1,q,t)L}\left(f(g)\right) - \varepsilon\right)\left(\log^{[q-1]} r \cdot \exp^{[t+1]} L(r)\right)^{\rho^{(m,q,t)L}(f(g))}, \tag{4.3.3}$$

$$\log^{[m-1]} T(r, f_u(g_v)) \geqslant \left(\overline{\sigma}^{(m,q,t)L}\left(f_u(g_v)\right) - \varepsilon\right)\left(\log^{[q-1]} r \cdot \exp^{[t+1]} L(r)\right)^{\rho^{(m,q,t)L}(f_u(g_v))}$$

$$i.e.,\ \log^{[m-1]} T(r, f_u(g_v)) \geqslant \left(\overline{\sigma}^{(m,q,t)L}\left(f(g)\right) - \varepsilon\right)\left(\log^{[q-1]} r \cdot \exp^{[t+1]} L(r)\right)^{\rho^{(m,q,t)L}(f(g))} \tag{4.3.4}$$

and

$$T(r, f_u) \leq \left(\sigma^{(1,q,t)L}\left(f_u\right) + \varepsilon\right)\left(\log^{[q-1]} r \cdot \exp^{[t+1]} L(r)\right)^{\rho^{(1,q,t)L}(f_u)}$$

$$i.e,\ T(r, f_u) \leq \left(u \cdot \sigma^{(1,q,t)L}\left(f\right) + \varepsilon\right)\left(\log^{[q-1]} r \cdot \exp^{[t+1]} L(r)\right)^{\rho^{(l,q,t)L}(f)}, \tag{4.3.5}$$

$$\log^{[l-1]} T(r, f_u) \leq \left(\sigma^{(l,q,t)L}\left(f_u\right) + \varepsilon\right)\left(\log^{[q-1]} r \cdot \exp^{[t+1]} L(r)\right)^{\rho^{(l,q,t)L}(f_u)}$$

$$i.e,\ \log^{[l-1]} T(r, f_u) \leq \left(\sigma^{(l,q,t)L}\left(f\right) + \varepsilon\right)\left(\log^{[q-1]} r \cdot \exp^{[t+1]} L(r)\right)^{\rho^{(l,q,t)L}(f)}. \tag{4.3.6}$$

Now from (4.3.3), (4.3.5) and the condition $\rho^{(1,q,t)L}\left(f(g)\right) = \rho^{(1,q,t)L}\left(f\right)$, it follows for all large values of r that

$$\frac{T(r, f_u(g_v))}{T(r, f_u)} \geqslant \frac{t \cdot \overline{\sigma}^{(1,q,t)L}\left(f(g)\right) - \varepsilon}{u \cdot \sigma^{(1,q,t)L}\left(f\right) + \varepsilon}.$$

As $\varepsilon\,(> 0)$ is arbitrary , we obtain that

$$\liminf_{r \to +\infty} \frac{T(r, f_u(g_v))}{T(r, f_u)} \geqslant \frac{t \cdot \overline{\sigma}^{(1,q,t)L}\left(f(g)\right)}{u \cdot \sigma^{(1,q,t)L}\left(f\right)}. \tag{4.3.7}$$

Further from (4.3.4), (4.3.6) and the condition $\rho^{(m,q,t)L}\left(f(g)\right) = \rho^{(l,q,t)L}\left(f\right)$, it follows for all large values of r that

$$\frac{\log^{[m-1]} T(r, f_u(g_v))}{\log^{[l-1]} T(r, f_u)} \geqslant \frac{\overline{\sigma}^{(l,q,t)L}\left(f(g)\right) - \varepsilon}{\sigma^{(l,q,t)L}\left(f\right) + \varepsilon}.$$

As $\varepsilon\,(>0)$ is arbitrary , we obtain that

$$\liminf_{r\to+\infty}\frac{\log^{[m-1]}T(r,f_u(g_v))}{\log^{[l-1]}T(r,f_u)}\geqslant\frac{\overline{\sigma}^{(l,q,t)L}\left(f(g)\right)}{\sigma^{(l,q,t)L}\left(f\right)}. \tag{4.3.8}$$

Again for a sequence of values of r tending to infinity,

$$T(r,f_u(g_v))\leq\left(\overline{\sigma}^{(1,q,t)L}\left(f_u(g_v)\right)+\varepsilon\right)\left(\log^{[q-1]}r\cdot\exp^{[t+1]}L\left(r\right)\right)^{\rho^{(1,q,t)L}(f_u(g_v))}$$

$$i.e.,\ T(r,f_u(g_v))\leq\left(t\cdot\ \overline{\sigma}^{(1,q,t)L}\left(f(g)\right)+\varepsilon\right)\left(\log^{[q-1]}r\cdot\exp^{[t+1]}L\left(r\right)\right)^{\rho^{(1,q,t)L}(f(g))}, \tag{4.3.9}$$

$$\log^{[m-1]}T(r,f_u(g_v))\leq\left(\overline{\sigma}^{(m,q,t)L}\left(f_u(g_v)\right)+\varepsilon\right)\left(\log^{[q-1]}r\cdot\exp^{[t+1]}L\left(r\right)\right)^{\rho^{(m,q,t)L}(f_u(g_v))}$$

$$i.e.,\ \log^{[m-1]}T(r,f_u(g_v))\leq\left(\overline{\sigma}^{(m,q,t)L}\left(f(g)\right)+\varepsilon\right)\left(\log^{[q-1]}r\cdot\exp^{[t+1]}L\left(r\right)\right)^{\rho^{(m,q,t)L}(f(g))} \tag{4.3.10}$$

and for all sufficiently large values of r,

$$T(r,f_u)\geqslant\left(\overline{\sigma}^{(1,q,t)L}\left(f_u\right)-\varepsilon\right)\left(\log^{[q-1]}r\cdot\exp^{[t+1]}L\left(r\right)\right)^{\rho^{(1,q,t)L}(f_u)}$$

$$i.e.,\ T(r,f_u)\geqslant\left(u\cdot\ \overline{\sigma}^{(1,q,t)L}\left(f\right)-\varepsilon\right)\left(\log^{[q-1]}r\cdot\exp^{[t+1]}L\left(r\right)\right)^{\rho^{(l,q,t)L}(f)}. \tag{4.3.11}$$

$$\log^{[l-1]}T(r,f_u)\geqslant\left(\overline{\sigma}^{(l,q,t)L}\left(f_u\right)-\varepsilon\right)\left(\log^{[q-1]}r\cdot\exp^{[t+1]}L\left(r\right)\right)^{\rho^{(l,q,t)L}(f_u)}$$

$$i.e.,\ \log^{[l-1]}T(r,f_u)\geqslant\left(\overline{\sigma}^{(m,q,t)L}\left(f\right)-\varepsilon\right)\left(\log^{[q-1]}r\cdot\exp^{[t+1]}L\left(r\right)\right)^{\rho^{(m,q,t)L}(f)}. \tag{4.3.12}$$

Combining the condition $\rho^{(1,q,t)L}\left(f(g)\right)=\rho^{(1,q,t)L}\left(f\right)$, (4.3.9) and (4.3.11) we get for a sequence of values of r tending to infinity that

$$\frac{T(r,f_u(g_v))}{T(r,f_u)}\leq\frac{t\cdot\ \overline{\sigma}^{(1,q,t)L}\left(f(g)\right)+\varepsilon}{u\cdot\ \overline{\sigma}^{(1,q,t)L}\left(f\right)-\varepsilon}.$$

Since $\varepsilon\,(>0)$ is arbitrary, it follows that

$$\liminf_{r\to+\infty}\frac{T(r,f_u(g_v))}{T(r,f_u)}\leq\frac{t\cdot\ \overline{\sigma}^{(1,q,t)L}\left(f(g)\right)}{u\cdot\ \overline{\sigma}^{(1,q,t)L}\left(f\right)}. \tag{4.3.13}$$

Also combining the condition $\rho^{(m,q,t)L}\left(f(g)\right)=\rho^{(l,q,t)L}\left(f\right)$, (4.3.10) and (4.3.12) we get for a sequence of values of r tending to infinity that

$$\frac{\log^{[m-1]}T(r,f_u(g_v))}{\log^{[l-1]}T(r,f_u)}\leq\frac{\overline{\sigma}^{(m,q,t)L}\left(f(g)\right)+\varepsilon}{\overline{\sigma}^{(l,q,t)L}\left(f\right)-\varepsilon}.$$

Since $\varepsilon\,(>0)$ is arbitrary, it follows that

$$\liminf_{r\to+\infty}\frac{\log^{[m-1]}T(r,f_u(g_v))}{\log^{[l-1]}T(r,f_u)}\leq\frac{\overline{\sigma}^{(m,q,t)L}(f(g))}{\overline{\sigma}^{(l,q,t)L}(f)}. \tag{4.3.14}$$

Also for a sequence of values of r tending to infinity that

$$T(r,f_u)\leq\left(\overline{\sigma}^{(1,q,t)L}(f_u)+\varepsilon\right)\left(\log^{[q-1]}r\cdot\exp^{[t+1]}L(r)\right)^{\rho^{(1,q,t)L}(f_u)}$$

$$i.e.,\ T(r,f_u)\leq\left(u\cdot\ \overline{\sigma}^{(1,q,t)L}(f)+\varepsilon\right)\left(\log^{[q-1]}r\cdot\exp^{[t+1]}L(r)\right)^{\rho^{(l,q,t)L}(f)}. \tag{4.3.15}$$

$$\log^{[l-1]}T(r,f_u)\leq\left(\overline{\sigma}^{(l,q,t)L}(f_u)+\varepsilon\right)\left(\log^{[q-1]}r\cdot\exp^{[t+1]}L(r)\right)^{\rho^{(l,q,t)L}(f_u)}$$

$$i.e.,\ \log^{[l-1]}T(r,f_u)\leq\left(\overline{\sigma}^{(l,q,t)L}(f)+\varepsilon\right)\left(\log^{[q-1]}r\cdot\exp^{[t+1]}L(r)\right)^{\rho^{(l,q,t)L}(f)}. \tag{4.3.16}$$

Now from (4.3.3), (4.3.15) and the condition $\rho^{(1,q,t)L}(f(g))=\rho^{(1,q,t)L}(f)$, we obtain for a sequence of values of r tending to infinity that

$$\frac{T(r,f_u(g_v))}{T(r,f_u)}\geq\frac{t\cdot\ \overline{\sigma}^{(1,q,t)L}(f(g))-\varepsilon}{u\cdot\ \overline{\sigma}^{(1,q,t)L}(f)+\varepsilon}.$$

As $\varepsilon\,(>0)$ is arbitrary, we get from above that

$$\limsup_{r\to+\infty}\frac{T(r,f_u(g_v))}{T(r,f_u)}\geq\frac{t\cdot\ \overline{\sigma}^{(1,q,t)L}(f(g))}{u\cdot\ \overline{\sigma}^{(1,q,t)L}(f)}. \tag{4.3.17}$$

Further from (4.3.4), (4.3.16) and the condition $\rho^{(m,q,t)L}(f(g))=\rho^{(l,q,t)L}(f)$, we obtain for a sequence of values of r tending to infinity that

$$\frac{\log^{[m-1]}T(r,f_u(g_v))}{\log^{[l-1]}T(r,f_u)}\geq\frac{\overline{\sigma}^{(m,q,t)L}(f(g))-\varepsilon}{\overline{\sigma}^{(l,q,t)L}(f)+\varepsilon}.$$

As $\varepsilon\,(>0)$ is arbitrary, we get from above that

$$\limsup_{r\to+\infty}\frac{\log^{[m-1]}T(r,f_u(g_v))}{\log^{[l-1]}T(r,f_u)}\geq\frac{\overline{\sigma}^{(m,q,t)L}(f(g))}{\overline{\sigma}^{(l,q,t)L}(f)}. \tag{4.3.18}$$

Also for all sufficiently large values of r,

$$T(r,f_u(g_v))\leq\left(\sigma^{(1,q,t)L}(f_u(g_v))+\varepsilon\right)\left(\log^{[q-1]}r\cdot\exp^{[t+1]}L(r)\right)^{\rho^{(1,q,t)L}(f_u(g_v))}$$

$$i.e.,\ T(r,f_u(g_v))\leq\left(t\cdot\ \sigma^{(1,q,t)L}(f(g))+\varepsilon\right)\left(\log^{[q-1]}r\cdot\exp^{[t+1]}L(r)\right)^{\rho^{(1,q,t)L}(f(g))}, \tag{4.3.19}$$

$$\log^{[m-1]} T(r, f_u(g_v)) \leq \left(\sigma^{(m,q,t)L}(f_u(g_v)) + \varepsilon\right)\left(\log^{[q-1]} r \cdot \exp^{[t+1]} L(r)\right)^{\rho^{(m,q,t)L}(f_u(g_v))}$$

$$i.e., \ \log^{[m-1]} T(r, f_u(g_v)) \leq \left(\sigma^{(m,q,t)L}(f(g)) + \varepsilon\right)\left(\log^{[q-1]} r \cdot \exp^{[t+1]} L(r)\right)^{\rho^{(m,q,t)L}(f(g))}. \tag{4.3.20}$$

As the condition $\rho^{(1,q,t)L}(f(g)) = \rho^{(1,q,t)L}(f)$, it follows from (4.3.11) and (4.3.19) for all large values of r that

$$\frac{T(r, f_u(g_v))}{T(r, f_u)} \leq \frac{t \cdot \sigma^{(1,q,t)L}(f(g)) + \varepsilon}{u \cdot \overline{\sigma}^{(1,q,t)L}(f) - \varepsilon}.$$

Since $\varepsilon\,(> 0)$ is arbitrary, we obtain that

$$\limsup_{r\to+\infty} \frac{T(r, f_u(g_v))}{T(r, f_u)} \leq \frac{t \cdot \sigma^{(1,q,t)L}(f(g))}{u \cdot \overline{\sigma}^{(1,q,t)L}(f)}. \tag{4.3.21}$$

Similarly as the condition $\rho^{(m,q,t)L}(f(g)) = \rho^{(l,q,t)L}(f)$, it follows from (4.3.12) and (4.3.20) for all large values of r that

$$\frac{\log^{[m-1]} T(r, f_u(g_v))}{\log^{[l-1]} T(r, f_u)} \leq \frac{\sigma^{(m,q,t)L}(f(g)) + \varepsilon}{\overline{\sigma}^{(l,q,t)L}(f) - \varepsilon}.$$

Since $\varepsilon\,(> 0)$ is arbitrary, we obtain that

$$\limsup_{r\to+\infty} \frac{\log^{[m-1]} T(r, f_u(g_v))}{\log^{[l-1]} T(r, f_u)} \leq \frac{\sigma^{(m,q,t)L}(f(g))}{\overline{\sigma}^{(l,q,t)L}(f)}. \tag{4.3.22}$$

Thus the first part of the theorem follows from (4.3.7), (4.3.13), (4.3.17) and (4.3.21). Similarly the second part of the theorem follows from (4.3.8), (4.3.14), (4.3.18) and (4.3.22).

Similarly we may state the following theorem without proof for the right factor g of the composite function $f(g)$:

Theorem 4.3.2 *Let $f(z)$ be a meromorphic function and $g(z)$ be a non constant entire function such that $0 < \overline{\sigma}^{(m,q,t)L}(f(g)) \leq \sigma^{(m,q,t)L}(f(g)) < \infty$, $0 < \overline{\sigma}^{(l,q,t)L}(g) \leq \sigma^{(l,q,t)L}(g) < \infty$, $\rho^{(m,q,t)L}(f(g)) = \rho^{(l,q,t)L}(g)$. Also let f_u and g_v be integer translations of $f(z)$ and $g(z)$, respectively, for $u, v \in \mathbb{N}$. If $f_u(g_v) = h_t$, where h is a meromorphic function and $t \in \mathbb{N}$, then*

$$\frac{t \cdot \overline{\sigma}^{(1,q,t)L}(f(g))}{v \cdot \sigma^{(1,q,t)L}(g)} \leq \liminf_{r\to+\infty} \frac{T(r, f_u(g_v))}{T(r, g_v)} \leq \frac{t \cdot \overline{\sigma}^{(1,q,t)L}(f(g))}{v \cdot \overline{\sigma}^{(1,q,t)L}(g)} \leq$$

$$\limsup_{r\to+\infty} \frac{T(r, f_u(g_v))}{T(r, g_v)} \leq \frac{t \cdot \sigma^{(1,q,t)L}(f(g))}{v \cdot \overline{\sigma}^{(1,q,t)L}(g)}$$

and

$$\frac{\overline{\sigma}^{(m,q,t)L}\left(f(g)\right)}{\sigma^{(l,q,t)L}\left(g\right)} \leq \liminf_{r\to+\infty}\frac{\log^{[m-1]}T\left(r,f_u(g_v)\right)}{\log^{[l-1]}T\left(r,g_v\right)} \leq \frac{\overline{\sigma}^{(m,q,t)L}\left(f(g)\right)}{\overline{\sigma}^{(l,q,t)L}\left(g\right)} \leq \limsup_{r\to+\infty}\frac{\log^{[m-1]}T\left(r,f_u(g_v)\right)}{\log^{[l-1]}T\left(r,g_v\right)} \leq \frac{\sigma^{(m,q,t)L}\left(f(g)\right)}{\overline{\sigma}^{(l,q,t)L}\left(g\right)}$$

for $m > 1$ *and* $l > 1$.

Theorem 4.3.3 *Let* $f\left(z\right)$ *be a meromorphic function and* $g\left(z\right)$ *be a non constant entire function such that* $0 < \sigma^{(m,q,t)L}\left(f(g)\right) < \infty$, $0 < \sigma^{(l,q,t)L}\left(f\right) < \infty$, $\rho^{(m,q,t)L}\left(f(g)\right) = \rho^{(l,q,t)L}\left(f\right)$. *Also let* f_u *and* g_v *be integer translations of* $f\left(z\right)$ *and* $g\left(z\right)$, *respectively, for* $u, v \in \mathbb{N}$. *If* $f_u(g_v) = h_t$, *where* h *is a meromorphic function and* $t \in \mathbb{N}$, *then*

$$\liminf_{r\to+\infty}\frac{T\left(r,f_u(g_v)\right)}{T\left(r,f_u\right)} \leq \frac{t\cdot\ \sigma^{(1,q,t)L}\left(f(g)\right)}{u\cdot\ \sigma^{(1,q,t)L}\left(f\right)} \leq \limsup_{r\to+\infty}\frac{T\left(r,f_u(g_v)\right)}{T\left(r,f_u\right)}$$

and

$$\liminf_{r\to+\infty}\frac{\log^{[m-1]}T\left(r,f_u(g_v)\right)}{\log^{[l-1]}T\left(r,f_u\right)} \leq \frac{\sigma^{(m,q,t)L}\left(f(g)\right)}{\sigma^{(l,q,t)L}\left(f\right)} \leq \limsup_{r\to+\infty}\frac{\log^{[m-1]}T\left(r,f_u(g_v)\right)}{\log^{[l-1]}T\left(r,f_u\right)}$$

for $m > 1$ *and* $l > 1$.

Proof In view of Lemma 2.2.2, Lemma 4.2.2, (4.3.1) and (2.3.1), we get for a sequence of values of r tending to infinity that

$$T(r,f_u) \geqslant \left(\sigma^{(1,q,t)L}\left(f_u\right)-\varepsilon\right)\left(\log^{[q-1]}r\cdot\exp^{[t+1]}L\left(r\right)\right)^{\rho^{(1,q,t)L}\left(f_u\right)}$$

$$i.e.,\ T(r,f_u) \geqslant \left(u\cdot\ \sigma^{(1,q,t)L}\left(f\right)-\varepsilon\right)\left(\log^{[q-1]}r\cdot\exp^{[t+1]}L\left(r\right)\right)^{\rho^{(1,q,t)L}\left(f\right)}, \tag{4.3.23}$$

$$\log^{[l-1]}T(r,f_u) \geqslant \left(\sigma^{(l,q,t)L}\left(f_u\right)-\varepsilon\right)\left(\log^{[q-1]}r\cdot\exp^{[t+1]}L\left(r\right)\right)^{\rho^{(l,q,t)L}\left(f_u\right)}$$

$$i.e.,\ \log^{[l-1]}T(r,f_u) \geqslant \left(\sigma^{(l,q,t)L}\left(f\right)-\varepsilon\right)\left(\log^{[q-1]}r\cdot\exp^{[t+1]}L\left(r\right)\right)^{\rho^{(l,q,t)L}\left(f\right)}. \tag{4.3.24}$$

Now from (4.3.19), (4.3.23) and the condition $\rho^{(1,q,t)L}\left(f(g)\right) = \rho^{(1,q,t)L}\left(f\right)$, it follows for a sequence of values of r tending to infinity that

$$\frac{T(r,f_u(g_v))}{T(r,f_u)} \leq \frac{t\cdot\ \sigma^{(1,q,t)L}\left(f(g)\right)+\varepsilon}{u\cdot\ \sigma^{(1,q,t)L}\left(f\right)-\varepsilon}.$$

As $\varepsilon\,(>0)$ is arbitrary, we obtain that

$$\liminf_{r\to+\infty}\frac{T(r,f_u(g_v))}{T(r,f_u)} \leq \frac{t\cdot\ \sigma^{(1,q,t)L}\left(f(g)\right)}{u\cdot\ \sigma^{(1,q,t)L}\left(f\right)}. \tag{4.3.25}$$

Further from (4.3.20), (4.3.24) and the condition $\rho^{(m,q,t)L}(f(g)) = \rho^{(l,q,t)L}(f)$, it follows for a sequence of values of r tending to infinity that

$$\frac{\log^{[m-1]} T(r, f_u(g_v))}{\log^{[l-1]} T(r, f_u)} \leq \frac{\sigma^{(m,q,t)L}(f(g)) + \varepsilon}{\sigma^{(l,q,t)L}(f) - \varepsilon}.$$

As $\varepsilon\,(> 0)$ is arbitrary, we obtain that

$$\liminf_{r\to+\infty} \frac{\log^{[m-1]} T(r, f_u(g_v))}{\log^{[l-1]} T(r, f_u)} \leq \frac{\sigma^{(m,q,t)L}(f(g))}{\sigma^{(l,q,t)L}(f)}. \tag{4.3.26}$$

Again for a sequence of values of r tending to infinity that

$$T(r, f_u(g_v)) \geqslant \left(\sigma^{(1,q,t)L}(f_u(g_v)) - \varepsilon\right)\left(\log^{[q-1]} r \cdot \exp^{[t+1]} L(r)\right)^{\rho^{(1,q,t)L}(f_u(g_v))}$$

$$i.e.,\ T(r, f_u(g_v)) \geqslant \left(t \cdot \sigma^{(1,q,t)L}(f(g)) - \varepsilon\right)\left(\log^{[q-1]} r \cdot \exp^{[t+1]} L(r)\right)^{\rho^{(1,q,t)L}(f(g))}. \tag{4.3.27}$$

$$\log^{[m-1]} T(r, f_u(g_v)) \geqslant \left(\sigma^{(m,q,t)L}(f_u(g_v)) - \varepsilon\right)\left(\log^{[q-1]} r \cdot \exp^{[t+1]} L(r)\right)^{\rho^{(m,q,t)L}(f_u(g_v))}$$

$$i.e.,\ \log^{[m-1]} T(r, f_u(g_v)) \geqslant \left(\sigma^{(m,q,t)L}(f(g)) - \varepsilon\right)\left(\log^{[q-1]} r \cdot \exp^{[t+1]} L(r)\right)^{\rho^{(m,q,t)L}(f(g))}. \tag{4.3.28}$$

So combining the condition $\rho^{(1,q,t)L}(f(g)) = \rho^{(1,q,t)L}(f)$, (4.3.5) and (4.3.27) we get for a sequence of values of r tending to infinity ,

$$\frac{T(r, f_u(g_v))}{T(r, f_u)} \geqslant \frac{t \cdot \sigma^{(1,q,t)L}(f(g)) - \varepsilon}{u \cdot \sigma^{(1,q,t)L}(f) + \varepsilon}.$$

Since $\varepsilon\,(> 0)$ is arbitrary, it follows that

$$\limsup_{r\to+\infty} \frac{T(r, f_u(g_v))}{T(r, f_u)} \geqslant \frac{t \cdot \sigma^{(1,q,t)L}(f(g))}{u \cdot \sigma^{(1,q,t)L}(f)}. \tag{4.3.29}$$

Similarly, combining the condition $\rho^{(m,q,t)L}(f(g)) = \rho^{(l,q,t)L}(f)$, (4.3.6) and (4.3.28) we get for a sequence of values of r tending to infinity ,

$$\frac{\log^{[m-1]} T(r, f_u(g_v))}{\log^{[l-1]} T(r, f_u)} \geqslant \frac{\sigma^{(m,q,t)L}(f(g)) - \varepsilon}{\sigma^{(l,q,t)L}(f) + \varepsilon}.$$

Since $\varepsilon\,(> 0)$ is arbitrary, it follows that

$$\limsup_{r\to+\infty} \frac{\log^{[m-1]} T(r, f_u(g_v))}{\log^{[l-1]} T(r, f_u)} \geqslant \frac{\sigma^{(m,q,t)L}(f(g))}{\sigma^{(l,q,t)L}(f)}. \tag{4.3.30}$$

Thus the first part of the theorem follows from (4.3.25) and (4.3.29). Similarly, the second part of the theorem follows from (4.3.26) and (4.3.30).

Theorem 4.3.4 *Let $f(z)$ be a meromorphic function and $g(z)$ be a non constant entire function such that $0 < \sigma^{(m,q,t)L}(f(g)) < \infty$, $0 < \sigma^{(l,q,t)L}(g) < \infty$, $\rho^{(m,q,t)L}(f(g)) = \rho^{(l,q,t)L}(g)$. Also let f_u and g_v be integer translations of $f(z)$ and $g(z)$, respectively, for $u, v \in \mathbb{N}$. If $f_u(g_v) = h_t$, where h is a meromorphic function and $t \in \mathbb{N}$, then*

$$\liminf_{r\to+\infty}\frac{T(r, f_u(g_v))}{T(r, g_v)} \leq \frac{t\cdot\ \sigma^{(1,q,t)L}(f(g))}{v\cdot\ \sigma^{(1,q,t)L}(g)} \leq \limsup_{r\to+\infty}\frac{T(r, f_u(g_v))}{T(r, g_v)}$$

and

$$\liminf_{r\to+\infty}\frac{\log^{[m-1]} T(r, f_u(g_v))}{\log^{[l-1]} T(r, g_v)} \leq \frac{\sigma^{(m,q,t)L}(f(g))}{\sigma^{(l,q,t)L}(g)} \leq \limsup_{r\to+\infty}\frac{\log^{[m-1]} T(r, f_u(g_v))}{\log^{[l-1]} T(r, g_v)}$$

for $m > 1$ and $l > 1$.

The proof is omitted.

The following theorem is a natural consequence of Theorem 4.3.1 and Theorem 4.3.3.

Theorem 4.3.5 *Let $f(z)$ be a meromorphic function and $g(z)$ be a non constant entire function such that $0 < \overline{\sigma}^{(m,q,t)L}(f(g)) \leq \sigma^{(m,q,t)L}(f(g)) < \infty$, $0 < \overline{\sigma}^{(l,q,t)L}(f) \leq \sigma^{(l,q,t)L}(f) < \infty$, $\rho^{(m,q,t)L}(f(g)) = \rho^{(l,q,t)L}(f)$. Also let f_u and g_v be integer translations of $f(z)$ and $g(z)$, respectively, for $u, v \in \mathbb{N}$. If $f_u(g_v) = h_t$, where h is a meromorphic function and $t \in \mathbb{N}$, then*

$$\liminf_{r\to+\infty}\frac{T(r, f_u(g_v))}{T(r, f_u)} \leq \min\left\{\frac{t\cdot\ \overline{\sigma}^{(1,q,t)L}(f(g))}{u\cdot\ \overline{\sigma}^{(1,q,t)L}(f)}, \frac{t\cdot\ \sigma^{(1,q,t)L}(f(g))}{u\cdot\ \sigma^{(1,q,t)L}(f)}\right\}$$
$$\leq \max\left\{\frac{t\cdot\ \overline{\sigma}^{(1,q,t)L}(f(g))}{u\cdot\ \overline{\sigma}^{(1,q,t)L}(f)}, \frac{t\cdot\ \sigma^{(1,q,t)L}(f(g))}{u\cdot\ \sigma^{(1,q,t)L}(f)}\right\} \leq \limsup_{r\to+\infty}\frac{T(r, f_u(g_v))}{T(r, f_u)}$$

and

$$\liminf_{r\to+\infty}\frac{\log^{[m-1]} T(r, f_u(g_v))}{\log^{[l-1]} T(r, f_u)} \leq \min\left\{\frac{\overline{\sigma}^{(m,q,t)L}(f(g))}{\overline{\sigma}^{(l,q,t)L}(f)}, \frac{\sigma^{(m,q,t)L}(f(g))}{\sigma^{(l,q,t)L}(f)}\right\}$$
$$\leq \max\left\{\frac{\overline{\sigma}^{(m,q,t)L}(f(g))}{\overline{\sigma}^{(l,q,t)L}(f)}, \frac{\sigma^{(m,q,t)L}(f(g))}{\sigma^{(l,q,t)L}(f)}\right\} \leq \limsup_{r\to+\infty}\frac{\log^{[m-1]} T(r, f_u(g_v))}{\log^{[l-1]} T(r, f_u)}$$

for $m > 1$ and $l > 1$.

The proof is omitted.

Combining Theorem 4.3.2 and Theorem 4.3.4 we may state the following theorem.

Theorem 4.3.6 *Let $f(z)$ be a meromorphic function and $g(z)$ be a non constant entire function such that $0 < \overline{\sigma}^{(m,q,t)L}(f(g)) \leq \sigma^{(m,q,t)L}(f(g)) < \infty$, $0 < \overline{\sigma}^{(l,q,t)L}(g) \leq \sigma^{(l,q,t)L}(g) < \infty$, $\rho^{(m,q,t)L}(f(g)) = \rho^{(l,q,t)L}(g)$. Also let f_u and g_v be integer translations of $f(z)$ and*

$g(z)$, respectively, for $u, v \in \mathbb{N}$. If $f_u(g_v) = h_t$, where h is a meromorphic function and $t \in \mathbb{N}$, then

$$\liminf_{r\to+\infty} \frac{T(r, f_u(g_v))}{T(r, g_v)} \leq \min\left\{\frac{t \cdot \overline{\sigma}^{(1,q,t)L}(f(g))}{v \cdot \overline{\sigma}^{(1,q,t)L}(g)}, \frac{t \cdot \sigma^{(1,q,t)L}(f(g))}{v \cdot \sigma^{(1,q,t)L}(g)}\right\}$$
$$\leq \max\left\{\frac{t \cdot \overline{\sigma}^{(1,q,t)L}(f(g))}{v \cdot \overline{\sigma}^{(1,q,t)L}(g)}, \frac{t \cdot \sigma^{(1,q,t)L}(f(g))}{v \cdot \sigma^{(1,q,t)L}(g)}\right\} \leq \limsup_{r\to+\infty} \frac{T(r, f_u(g_v))}{T(r, g_v)}$$

and

$$\liminf_{r\to+\infty} \frac{\log^{[m-1]} T(r, f_u(g_v))}{\log^{[l-1]} T(r, g_v)} \leq \min\left\{\frac{\overline{\sigma}^{(m,q,t)L}(f(g))}{\overline{\sigma}^{(l,q,t)L}(g)}, \frac{\sigma^{(m,q,t)L}(f(g))}{\sigma^{(l,q,t)L}(g)}\right\}$$
$$\leq \max\left\{\frac{\overline{\sigma}^{(m,q,t)L}(f(g))}{\overline{\sigma}^{(l,q,t)L}(g)}, \frac{\sigma^{(m,q,t)L}(f(g))}{\sigma^{(l,q,t)L}(g)}\right\} \leq \limsup_{r\to+\infty} \frac{\log^{[m-1]} T(r, f_u(g_v))}{\log^{[l-1]} T(r, g_v)}$$

for $m > 1$ and $l > 1$.

Now in the line of Theorem 4.3.1 to Theorem 4.3.6, respectively, one can easily prove the following six theorems using the notion of L^*weak type, and therefore their proofs are omitted.

Theorem 4.3.7 *Let $f(z)$ be a meromorphic function and $g(z)$ be a non constant entire function such that $0 < \overline{\tau}^{(m,q,t)L}(f(g)) \leq \tau^{(m,q,t)L}(f(g)) < \infty$, $0 < \overline{\tau}^{(l,q,t)L}(f) \leq \tau^{(l,q,t)L}(f) < \infty$, $\lambda^{(m,q,t)L}(f(g)) = \lambda^{(l,q,t)L}(f)$. Also let f_u and g_v be integer translations of $f(z)$ and $g(z)$, respectively, for $u, v \in \mathbb{N}$. If $f_u(g_v) = h_t$, where h is a meromorphic function and $t \in \mathbb{N}$, then*

$$\frac{t \cdot \overline{\tau}^{(1,q,t)L}(f(g))}{u \cdot \tau^{(1,q,t)L}(f)} \leq \liminf_{r\to+\infty} \frac{T(r, f_u(g_v))}{T(r, f_u)} \leq \frac{t \cdot \overline{\tau}^{(1,q,t)L}(f(g))}{u \cdot \overline{\tau}^{(1,q,t)L}(f)}$$
$$\leq \limsup_{r\to+\infty} \frac{T(r, f_u(g_v))}{T(r, f_u)} \leq \frac{t \cdot \tau^{(1,q,t)L}(f(g))}{u \cdot \overline{\tau}^{(1,q,t)L}(f)}$$

and

$$\frac{\overline{\tau}^{(m,q,t)L}(f(g))}{\tau^{(l,q,t)L}(f)} \leq \liminf_{r\to+\infty} \frac{\log^{[m-1]} T(r, f_u(g_v))}{\log^{[l-1]} T(r, f_u)} \leq \frac{\overline{\tau}^{(m,q,t)L}(f(g))}{\overline{\tau}^{(l,q,t)L}(f)}$$
$$\leq \limsup_{r\to+\infty} \frac{\log^{[m-1]} T(r, f_u(g_v))}{\log^{[l-1]} T(r, f_u)} \leq \frac{\tau^{(m,q,t)L}(f(g))}{\overline{\tau}^{(l,q,t)L}(f)}$$

for $m > 1$ and $l > 1$.

Theorem 4.3.8 *Let $f(z)$ be a meromorphic function and $g(z)$ be a non constant entire function such that $0 < \tau^{(m,q,t)L}(f(g)) < \infty$, $0 < \tau^{(l,q,t)L}(f) < \infty$, $\lambda^{(m,q,t)L}(f(g)) =$*

$\lambda^{(l,q,t)L}(f)$. *Also let* f_u *and* g_v *be integer translations of* $f(z)$ *and* $g(z)$, *respectively, for* $u, v \in \mathbb{N}$. *If* $f_u(g_v) = h_t$, *where* h *is a meromorphic function and* $t \in \mathbb{N}$, *then*

$$\liminf_{r \to +\infty} \frac{T(r, f_u(g_v))}{T(r, f_u)} \leq \frac{t \cdot \tau^{(1,q,t)L}(f(g))}{u \cdot \tau^{(1,q,t)L}(f)} \leq \limsup_{r \to +\infty} \frac{T(r, f_u(g_v))}{T(r, f_u)}$$

and

$$\liminf_{r \to +\infty} \frac{\log^{[m-1]} T(r, f_u(g_v))}{\log^{[l-1]} T(r, f_u)} \leq \frac{\tau^{(m,q,t)L}(f(g))}{\tau^{(l,q,t)L}(f)} \leq \limsup_{r \to +\infty} \frac{\log^{[m-1]} T(r, f_u(g_v))}{\log^{[l-1]} T(r, f_u)}$$

for $m > 1$ *and* $l > 1$.

Theorem 4.3.9 *Let* $f(z)$ *be a meromorphic function and* $g(z)$ *be a non constant entire function such that* $0 < \overline{\tau}^{(m,q,t)L}(f(g)) \leq \tau^{(m,q,t)L}(f(g)) < \infty$, $0 < \overline{\tau}^{(l,q,t)L}(f) \leq \tau^{(l,q,t)L}(f) < \infty$, $\lambda^{(m,q,t)L}(f(g)) = \lambda^{(l,q,t)L}(f)$. *Also let* f_u *and* g_v *be integer translations of* $f(z)$ *and* $g(z)$, *respectively, for* $u, v \in \mathbb{N}$. *If* $f_u(g_v) = h_t$, *where* h *is a meromorphic function and* $t \in \mathbb{N}$, *then*

$$\liminf_{r \to +\infty} \frac{T(r, f_u(g_v))}{T(r, f_u)} \leq \min\left\{\frac{t \cdot \overline{\tau}^{(1,q,t)L}(f(g))}{u \cdot \overline{\tau}^{(1,q,t)L}(f)}, \frac{t \cdot \tau^{(1,q,t)L}(f(g))}{u \cdot \tau^{(1,q,t)L}(f)}\right\}$$
$$\leq \max\left\{\frac{t \cdot \overline{\tau}^{(1,q,t)L}(f(g))}{u \cdot \overline{\tau}^{(1,q,t)L}(f)}, \frac{t \cdot \tau^{(1,q,t)L}(f(g))}{u \cdot \tau^{(1,q,t)L}(f)}\right\} \leq \limsup_{r \to +\infty} \frac{T(r, f_u(g_v))}{T(r, f_u)}$$

and

$$\liminf_{r \to +\infty} \frac{\log^{[m-1]} T(r, f_u(g_v))}{\log^{[l-1]} T(r, f_u)} \leq \min\left\{\frac{\overline{\tau}^{(m,q,t)L}(f(g))}{\overline{\tau}^{(l,q,t)L}(f)}, \frac{\tau^{(m,q,t)L}(f(g))}{\tau^{(l,q,t)L}(f)}\right\}$$
$$\leq \max\left\{\frac{\overline{\tau}^{(m,q,t)L}(f(g))}{\overline{\tau}^{(l,q,t)L}(f)}, \frac{\tau^{(m,q,t)L}(f(g))}{\tau^{(l,q,t)L}(f)}\right\} \leq \limsup_{r \to +\infty} \frac{\log^{[m-1]} T(r, f_u(g_v))}{\log^{[l-1]} T(r, f_u)}$$

for $m > 1$ *and* $l > 1$.

Theorem 4.3.10 *Let* $f(z)$ *be a meromorphic function and* $g(z)$ *be a non constant entire function such that* $0 < \overline{\tau}^{(m,q,t)L}(f(g)) \leq \tau^{(m,q,t)L}(f(g)) < \infty$, $0 < \overline{\tau}^{(l,q,t)L}(g) \leq \tau^{(l,q,t)L}(g) < \infty$, $\lambda^{(m,q,t)L}(f(g)) = \lambda^{(l,q,t)L}(g)$. *Also let* f_u *and* g_v *be integer translations of* $f(z)$ *and* $g(z)$, *respectively, for* $u, v \in \mathbb{N}$. *If* $f_u(g_v) = h_t$, *where* h *is a meromorphic function and* $t \in \mathbb{N}$, *then*

$$\frac{t \cdot \overline{\tau}^{(1,q,t)L}(f(g))}{v \cdot \tau^{(1,q,t)L}(g)} \leq \liminf_{r \to +\infty} \frac{T(r, f_u(g_v))}{T(r, g_v)} \leq \frac{t \cdot \overline{\tau}^{(1,q,t)L}(f(g))}{v \cdot \overline{\tau}^{(1,q,t)L}(g)}$$
$$\leq \limsup_{r \to +\infty} \frac{T(r, f_u(g_v))}{T(r, g_v)} \leq \frac{t \cdot \tau^{(1,q,t)L}(f(g))}{v \cdot \overline{\tau}^{(1,q,t)L}(g)}$$

and

$$\frac{\overline{\tau}^{(m,q,t)L}\left(f(g)\right)}{\tau^{(l,q,t)L}\left(g\right)} \leq \liminf_{r\to+\infty} \frac{\log^{[m-1]} T\left(r, f_u(g_v)\right)}{\log^{[l-1]} T\left(r, g_v\right)} \leq \frac{\overline{\tau}^{(m,q,t)L}\left(f(g)\right)}{\overline{\tau}^{(l,q,t)L}\left(g\right)}$$
$$\leq \limsup_{r\to+\infty} \frac{\log^{[m-1]} T\left(r, f_u(g_v)\right)}{\log^{[l-1]} T\left(r, g_v\right)} \leq \frac{\tau^{(m,q,t)L}\left(f(g)\right)}{\overline{\tau}^{(l,q,t)L}\left(g\right)}$$

for $m > 1$ *and* $l > 1$.

Theorem 4.3.11 *Let* $f(z)$ *be a meromorphic function and* $g(z)$ *be a non constant entire function such that* $0 < \tau^{(m,q,t)L}(f(g)) < \infty$, $0 < \tau^{(l,q,t)L}(g) < \infty$, $\lambda^{(m,q,t)L}(f(g)) = \lambda^{(l,q,t)L}(g)$. *Also let* f_u *and* g_v *be integer translations of* $f(z)$ *and* $g(z)$, *respectively, for* $u, v \in \mathbb{N}$. *If* $f_u(g_v) = h_t$, *where* h *is a meromorphic function and* $t \in \mathbb{N}$, *then*

$$\liminf_{r\to+\infty} \frac{T\left(r, f_u(g_v)\right)}{T\left(r, g_v\right)} \leq \frac{t \cdot \tau^{(1,q,t)L}\left(f(g)\right)}{v \cdot \tau^{(1,q,t)L}\left(g\right)} \leq \limsup_{r\to+\infty} \frac{T\left(r, f_u(g_v)\right)}{T\left(r, g_v\right)}$$

and

$$\liminf_{r\to+\infty} \frac{\log^{[m-1]} T\left(r, f_u(g_v)\right)}{\log^{[l-1]} T\left(r, g_v\right)} \leq \frac{\tau^{(m,q,t)L}\left(f(g)\right)}{\tau^{(l,q,t)L}\left(g\right)} \leq \limsup_{r\to+\infty} \frac{\log^{[m-1]} T\left(r, f_u(g_v)\right)}{\log^{[l-1]} T\left(r, g_v\right)}$$

for $m > 1$ *and* $l > 1$.

Theorem 4.3.12 *Let* $f(z)$ *be a meromorphic function and* $g(z)$ *be a non constant entire function such that* $0 < \overline{\tau}^{(m,q,t)L}(f(g)) \leq \tau^{(m,q,t)L}(f(g)) < \infty$, $0 < \overline{\tau}^{(l,q,t)L}(g) \leq \tau^{(l,q,t)L}(g) < \infty$, $\lambda^{(m,q,t)L}(f(g)) = \lambda^{(l,q,t)L}(g)$. *Also let* f_u *and* g_v *be integer translations of* $f(z)$ *and* $g(z)$, *respectively, for* $u, v \in \mathbb{N}$. *If* $f_u(g_v) = h_t$, *where* h *is a meromorphic function and* $t \in \mathbb{N}$, *then*

$$\liminf_{r\to+\infty} \frac{T\left(r, f_u(g_v)\right)}{T\left(r, g_v\right)} \leq \min\left\{\frac{t \cdot \overline{\tau}^{(1,q,t)L}\left(f(g)\right)}{v \cdot \overline{\tau}^{(1,q,t)L}\left(g\right)}, \frac{t \cdot \tau^{(1,q,t)L}\left(f(g)\right)}{v \cdot \tau^{(1,q,t)L}\left(g\right)}\right\}$$
$$\leq \max\left\{\frac{t \cdot \overline{\tau}^{(1,q,t)L}\left(f(g)\right)}{v \cdot \overline{\tau}^{(1,q,t)L}\left(g\right)}, \frac{t \cdot \tau^{(1,q,t)L}\left(f(g)\right)}{v \cdot \tau^{(1,q,t)L}\left(g\right)}\right\} \leq \limsup_{r\to+\infty} \frac{T\left(r, f_u(g_v)\right)}{T\left(r, g_v\right)}$$

and

$$\liminf_{r\to+\infty} \frac{\log^{[m-1]} T\left(r, f_u(g_v)\right)}{\log^{[l-1]} T\left(r, g_v\right)} \leq \min\left\{\frac{\overline{\tau}^{(m,q,t)L}\left(f(g)\right)}{\overline{\tau}^{(l,q,t)L}\left(g\right)}, \frac{\tau^{(m,q,t)L}\left(f(g)\right)}{\tau^{(l,q,t)L}\left(g\right)}\right\}$$
$$\leq \max\left\{\frac{\overline{\tau}^{(m,q,t)L}\left(f(g)\right)}{\overline{\tau}^{(l,q,t)L}\left(g\right)}, \frac{\tau^{(m,q,t)L}\left(f(g)\right)}{\tau^{(l,q,t)L}\left(g\right)}\right\} \leq \limsup_{r\to+\infty} \frac{\log^{[m-1]} T\left(r, f_u(g_v)\right)}{\log^{[l-1]} T\left(r, g_v\right)}$$

for $m > 1$ *and* $l > 1$.

We may now state the following theorems without their proofs based on $(p, q, t)L$-th type and $(p, q, t)L$-th weak type of entire and meromorphic functions.

Theorem 4.3.13 *Let $f(z)$ be a meromorphic function and $g(z)$ be a non constant entire function such that* $0 < \overline{\sigma}^{(m,q,t)L}(f(g)) \leq \sigma^{(m,q,t)L}(f(g)) < \infty$, $0 < \overline{\tau}^{(l,q,t)L}(f) \leq \tau^{(l,q,t)L}(f) < \infty$, $\rho^{(m,q,t)L}(f(g)) = \lambda^{(l,q,t)L}(f)$. *Also let f_u and g_v be integer translations of $f(z)$ and $g(z)$, respectively, for $u, v \in \mathbb{N}$. If $f_u(g_v) = h_t$, where h is a meromorphic function and $t \in \mathbb{N}$, then*

$$\frac{t \cdot \overline{\sigma}^{(1,q,t)L}(f(g))}{u \cdot \tau^{(1,q,t)L}(f)} \leq \liminf_{r\to+\infty} \frac{T(r, f_u(g_v))}{T(r, f_u)} \leq \frac{t \cdot \overline{\sigma}^{(1,q,t)L}(f(g))}{u \cdot \overline{\tau}^{(1,q,t)L}(f)}$$
$$\leq \limsup_{r\to+\infty} \frac{T(r, f_u(g_v))}{T(r, f_u)} \leq \frac{t \cdot \sigma^{(1,q,t)L}(f(g))}{u \cdot \overline{\tau}^{(1,q,t)L}(f)}$$

and

$$\frac{\overline{\sigma}^{(m,q,t)L}(f(g))}{\tau^{(l,q,t)L}(f)} \leq \liminf_{r\to+\infty} \frac{\log^{[m-1]} T(r, f_u(g_v))}{\log^{[l-1]} T(r, f_u)} \leq \frac{\overline{\sigma}^{(m,q,t)L}(f(g))}{\overline{\tau}^{(l,q,t)L}(f)}$$
$$\leq \limsup_{r\to+\infty} \frac{\log^{[m-1]} T(r, f_u(g_v))}{\log^{[l-1]} T(r, f_u)} \leq \frac{\sigma^{(m,q,t)L}(f(g))}{\overline{\tau}^{(l,q,t)L}(f)}$$

for $m > 1$ and $l > 1$.

Theorem 4.3.14 *Let $f(z)$ be a meromorphic function and $g(z)$ be a non constant entire function such that* $0 < \sigma^{(m,q,t)L}(f(g)) < \infty$, $0 < \sigma^{(l,q,t)L}(f) < \infty$, $\rho^{(m,q,t)L}(f(g)) = \lambda^{(l,q,t)L}(f)$. *Also let f_u and g_v be integer translations of $f(z)$ and $g(z)$, respectively, for $u, v \in \mathbb{N}$. If $f_u(g_v) = h_t$, where h is a meromorphic function and $t \in \mathbb{N}$, then*

$$\liminf_{r\to+\infty} \frac{T(r, f_u(g_v))}{T(r, f_u)} \leq \frac{t \cdot \sigma^{(1,q,t)L}(f(g))}{u \cdot \tau^{(1,q,t)L}(f)} \leq \limsup_{r\to+\infty} \frac{T(r, f_u(g_v))}{T(r, f_u)}$$

and

$$\liminf_{r\to+\infty} \frac{\log^{[m-1]} T(r, f_u(g_v))}{\log^{[l-1]} T(r, f_u)} \leq \frac{\sigma^{(m,q,t)L}(f(g))}{\tau^{(l,q,t)L}(f)} \leq \limsup_{r\to+\infty} \frac{\log^{[m-1]} T(r, f_u(g_v))}{\log^{[l-1]} T(r, f_u)}$$

for $m > 1$ and $l > 1$.

Theorem 4.3.15 *Let $f(z)$ be a meromorphic function and $g(z)$ be a non constant entire function such that* $0 < \overline{\sigma}^{(m,q,t)L}(f(g)) \leq \sigma^{(m,q,t)L}(f(g)) < \infty$, $0 < \overline{\tau}^{(l,q,t)L}(f) \leq \tau^{(l,q,t)L}(f) < \infty$, $\rho^{(m,q,t)L}(f(g)) = \lambda^{(l,q,t)L}(f)$. *Also let f_u and g_v be integer translations of $f(z)$ and $g(z)$, respectively, for $u, v \in \mathbb{N}$. If $f_u(g_v) = h_t$, where h is a meromorphic function and $t \in \mathbb{N}$, then*

$$\liminf_{r\to+\infty} \frac{T(r, f_u(g_v))}{T(r, f_u)} \leq \min\left\{\frac{t \cdot \overline{\sigma}^{(1,q,t)L}(f(g))}{u \cdot \overline{\tau}^{(1,q,t)L}(f)}, \frac{t \cdot \sigma^{(1,q,t)L}(f(g))}{u \cdot \tau^{(1,q,t)L}(f)}\right\}$$
$$\leq \max\left\{\frac{t \cdot \overline{\sigma}^{(1,q,t)L}(f(g))}{u \cdot \overline{\tau}^{(1,q,t)L}(f)}, \frac{t \cdot \sigma^{(1,q,t)L}(f(g))}{u \cdot \tau^{(1,q,t)L}(f)}\right\} \leq \limsup_{r\to+\infty} \frac{T(r, f_u(g_v))}{T(r, f_u)}$$

and

$$\liminf_{r\to+\infty}\frac{\log^{[m-1]}T(r,f_u(g_v))}{\log^{[l-1]}T(r,f_u)}\leq\min\left\{\frac{\overline{\sigma}^{(m,q,t)L}(f(g))}{\overline{\tau}^{(l,q,t)L}(f)},\frac{\sigma^{(m,q,t)L}(f(g))}{\tau^{(l,q,t)L}(f)}\right\}$$
$$\leq\max\left\{\frac{\overline{\sigma}^{(m,q,t)L}(f(g))}{\overline{\tau}^{(l,q,t)L}(f)},\frac{\sigma^{(m,q,t)L}(f(g))}{\tau^{(l,q,t)L}(f)}\right\}\leq\limsup_{r\to+\infty}\frac{\log^{[m-1]}T(r,f_u(g_v))}{\log^{[l-1]}T(r,f_u)}$$

for $m>1$ *and* $l>1$.

Theorem 4.3.16 *Let* $f(z)$ *be a meromorphic function and* $g(z)$ *be a non constant entire function such that* $0<\overline{\tau}^{(m,q,t)L}(f(g))\leq\tau^{(m,q,t)L}(f(g))<\infty$, $0<\overline{\sigma}^{(l,q,t)L}(f)\leq\sigma^{(l,q,t)L}(f)<\infty$, $\lambda^{(m,q,t)L}(f(g))=\rho^{(l,q,t)L}(f)$. *Also let* f_u *and* g_v *be integer translations of* $f(z)$ *and* $g(z)$, *respectively, for* $u,v\in\mathbb{N}$. *If* $f_u(g_v)=h_t$, *where* h *is a meromorphic function and* $t\in\mathbb{N}$, *then*

$$\frac{\overline{\tau}^{(1,q,t)L}(f(g))}{u\cdot\sigma^{(1,q,t)L}(f)}\leq\liminf_{r\to+\infty}\frac{T(r,f_u(g_v))}{T(r,f_u)}\leq\frac{t\cdot\overline{\tau}^{(1,q,t)L}(f(g))}{u\cdot\overline{\sigma}^{(1,q,t)L}(f)}$$
$$\leq\limsup_{r\to+\infty}\frac{T(r,f_u(g_v))}{T(r,f_u)}\leq\frac{t\cdot\tau^{(1,q,t)L}(f(g))}{u\cdot\overline{\sigma}^{(1,q,t)L}(f)}$$

and

$$\frac{\overline{\tau}^{(m,q,t)L}(f(g))}{\sigma^{(l,q,t)L}(f)}\leq\liminf_{r\to+\infty}\frac{\log^{[m-1]}T(r,f_u(g_v))}{\log^{[l-1]}T(r,f_u)}\leq\frac{\overline{\tau}^{(m,q,t)L}(f(g))}{\overline{\sigma}^{(l,q,t)L}(f)}$$
$$\leq\limsup_{r\to+\infty}\frac{\log^{[m-1]}T(r,f_u(g_v))}{\log^{[l-1]}T(r,f_u)}\leq\frac{\tau^{(m,q,t)L}(f(g))}{\overline{\sigma}^{(l,q,t)L}(f)}$$

for $m>1$ *and* $l>1$.

Theorem 4.3.17 *Let* $f(z)$ *be a meromorphic function and* $g(z)$ *be a non constant entire function such that* $0<\tau^{(m,q,t)L}(f(g))<\infty$, $0<\sigma^{(l,q,t)L}(f)<\infty$, $\lambda^{(m,q,t)L}(f(g))=\rho^{(l,q,t)L}(f)$. *Also let* f_u *and* g_v *be integer translations of* $f(z)$ *and* $g(z)$, *respectively, for* $u,v\in\mathbb{N}$. *If* $f_u(g_v)=h_t$, *where* h *is a meromorphic function and* $t\in\mathbb{N}$, *then*

$$\liminf_{r\to+\infty}\frac{T(r,f_u(g_v))}{T(r,f_u)}\leq\frac{t\cdot\tau^{(1,q,t)L}(f(g))}{u\cdot\sigma^{(1,q,t)L}(f)}\leq\limsup_{r\to+\infty}\frac{T(r,f_u(g_v))}{T(r,f_u)}$$

and

$$\liminf_{r\to+\infty}\frac{\log^{[m-1]}T(r,f_u(g_v))}{\log^{[l-1]}T(r,f_u)}\leq\frac{\tau^{(m,q,t)L}(f(g))}{\sigma^{(l,q,t)L}(f)}\leq\limsup_{r\to+\infty}\frac{\log^{[m-1]}T(r,f_u(g_v))}{\log^{[l-1]}T(r,f_u)}$$

for $m>1$ *and* $l>1$.

Theorem 4.3.18 *Let $f(z)$ be a meromorphic function and $g(z)$ be a non constant entire function such that $0 < \overline{\tau}^{(m,q,t)L}(f(g)) \leq \tau^{(m,q,t)L}(f(g)) < \infty$, $0 < \overline{\sigma}^{(l,q,t)L}(f) \leq \sigma^{(l,q,t)L}(f) < \infty$, $\lambda^{(m,q,t)L}(f(g)) = \rho^{(l,q,t)L}(f)$. Also let f_u and g_v be integer translations of $f(z)$ and $g(z)$, respectively, for $u, v \in \mathbb{N}$. If $f_u(g_v) = h_t$, where h is a meromorphic function and $t \in \mathbb{N}$, then*

$$\liminf_{r\to+\infty}\frac{T(r, f_u(g_v))}{T(r, f_u)} \leq \min\left\{\frac{t\cdot\ \overline{\tau}^{(1,q,t)L}(f(g))}{u\cdot\ \overline{\sigma}^{(1,q,t)L}(f)}, \frac{t\cdot\ \tau^{(1,q,t)L}(f(g))}{u\cdot\ \sigma^{(1,q,t)L}(f)}\right\}$$
$$\leq \max\left\{\frac{t\cdot\ \overline{\tau}^{(1,q,t)L}(f(g))}{u\cdot\ \overline{\sigma}^{(1,q,t)L}(f)}, \frac{t\cdot\ \tau^{(1,q,t)L}(f(g))}{u\cdot\ \sigma^{(1,q,t)L}(f)}\right\} \leq \limsup_{r\to+\infty}\frac{T(r, f_u(g_v))}{T(r, f_u)}$$

and

$$\liminf_{r\to+\infty}\frac{\log^{[m-1]} T(r, f_u(g_v))}{\log^{[l-1]} T(r, f_u)} \leq \min\left\{\frac{\overline{\tau}^{(m,q,t)L}(f(g))}{\overline{\sigma}^{(l,q,t)L}(f)}, \frac{\tau^{(m,q,t)L}(f(g))}{\sigma^{(l,q,t)L}(f)}\right\}$$
$$\leq \max\left\{\frac{\overline{\tau}^{(m,q,t)L}(f(g))}{\overline{\sigma}^{(l,q,t)L}(f)}, \frac{\tau^{(m,q,t)L}(f(g))}{\sigma^{(l,q,t)L}(f)}\right\} \leq \limsup_{r\to+\infty}\frac{\log^{[m-1]} T(r, f_u(g_v))}{\log^{[l-1]} T(r, f_u)}$$

for $m > 1$ and $l > 1$.

Theorem 4.3.19 *Let $f(z)$ be a meromorphic function and $g(z)$ be a non constant entire function such that $0 < \overline{\sigma}^{(1\backslash m,q,t)L}(f(g)) \leq \sigma^{(m,q,t)L}(f(g)) < \infty$, $0 < \overline{\tau}^{(l,q,t)L}(g) \leq \tau^{(l,q,t)L}(g) < \infty$, $\rho^{(m,q,t)L}(f(g)) = \lambda^{(l,q,t)L}(g)$. Also let f_u and g_v be integer translations of $f(z)$ and $g(z)$, respectively, for $u, v \in \mathbb{N}$. If $f_u(g_v) = h_t$, where h is a meromorphic function and $t \in \mathbb{N}$, then*

$$\frac{t\cdot\ \overline{\sigma}^{(1,q,t)L}(f(g))}{v\cdot\ \tau^{(1,q,t)L}(g)} \leq \liminf_{r\to+\infty}\frac{T(r, f_u(g_v))}{T(r, g_v)} \leq \frac{t\cdot\ \overline{\sigma}^{(1,q,t)L}(f(g))}{v\cdot\ \overline{\tau}^{(1,q,t)L}(g)}$$
$$\leq \limsup_{r\to+\infty}\frac{T(r, f_u(g_v))}{T(r, g_v)} \leq \frac{t\cdot\ \sigma^{(1,q,t)L}(f(g))}{v\cdot\ \overline{\tau}^{(1,q,t)L}(g)}$$

and

$$\frac{\overline{\sigma}^{(m,q,t)L}(f(g))}{\tau^{(l,q,t)L}(g)} \leq \liminf_{r\to+\infty}\frac{\log^{[m-1]} T(r, f_u(g_v))}{\log^{[l-1]} T(r, g_v)} \leq \frac{\overline{\sigma}^{(m,q,t)L}(f(g))}{\overline{\tau}^{(l,q,t)L}(g)}$$
$$\leq \limsup_{r\to+\infty}\frac{\log^{[m-1]} T(r, f_u(g_v))}{\log^{[l-1]} T(r, g_v)} \leq \frac{\sigma^{(m,q,t)L}(f(g))}{\overline{\tau}^{(l,q,t)L}(g)}$$

for $m > 1$ and $l > 1$.

Theorem 4.3.20 *Let $f(z)$ be a meromorphic function and $g(z)$ be a non constant entire function such that $0 < \sigma^{(m,q,t)L}(f(g)) < \infty$, $0 < \tau^{(l,q,t)L}(g) < \infty$, $\rho^{(m,q,t)L}(f(g)) =$*

$\lambda^{(l,q,t)L}(g)$. *Also let* f_u *and* g_v *be integer translations of* $f(z)$ *and* $g(z)$*, respectively, for* $u, v \in \mathbb{N}$. *If* $f_u(g_v) = h_t$*, where* h *is a meromorphic function and* $t \in \mathbb{N}$*, then*

$$\liminf_{r\to+\infty}\frac{T(r,f_u(g_v))}{T(r,g_v)} \le \frac{t\cdot\ \sigma^{(1,q,t)L}(f(g))}{v\cdot\ \tau^{(1,q,t)L}(g)} \le \limsup_{r\to+\infty}\frac{T(r,f_u(g_v))}{T(r,g_v)}$$

and

$$\liminf_{r\to+\infty}\frac{\log^{[m-1]}T(r,f_u(g_v))}{\log^{[l-1]}T(r,g_v)} \le \frac{\sigma^{(m,q,t)L}(f(g))}{\tau^{(l,q,t)L}(g)} \le \limsup_{r\to+\infty}\frac{\log^{[m-1]}T(r,f_u(g_v))}{\log^{[l-1]}T(r,g_v)}$$

for $m > 1$ *and* $l > 1$.

Theorem 4.3.21 *Let* $f(z)$ *be a meromorphic function and* $g(z)$ *be a non constant entire function such that* $0 < \overline{\sigma}^{(m,q,t)L}(f(g)) \le \sigma^{(m,q,t)L}(f(g)) < \infty$, $0 < \overline{\tau}^{(l,q,t)L}(g) \le \tau^{(l,q,t)L}(g) < \infty$, $\rho^{(m,q,t)L}(f(g)) = \lambda^{(l,q,t)L}(g)$. *Also let* f_u *and* g_v *be integer translations of* $f(z)$ *and* $g(z)$*, respectively, for* $u, v \in \mathbb{N}$. *If* $f_u(g_v) = h_t$*, where* h *is a meromorphic function and* $t \in \mathbb{N}$*, then*

$$\liminf_{r\to+\infty}\frac{T(r,f_u(g_v))}{T(r,g_v)} \le \min\left\{\frac{t\cdot\ \overline{\sigma}^{(1,q,t)L}(f(g))}{v\cdot\ \overline{\tau}^{(1,q,t)L}(g)}, \frac{t\cdot\ \sigma^{(1,q,t)L}(f(g))}{v\cdot\ \tau^{(1,q,t)L}(g)}\right\}$$
$$\le \max\left\{\frac{t\cdot\ \overline{\sigma}^{(1,q,t)L}(f(g))}{v\cdot\ \overline{\tau}^{(1,q,t)L}(g)}, \frac{t\cdot\ \sigma^{(1,q,t)L}(f(g))}{v\cdot\ \tau^{(1,q,t)L}(g)}\right\} \le \limsup_{r\to+\infty}\frac{T(r,f_u(g_v))}{T(r,g_v)}$$

and

$$\liminf_{r\to+\infty}\frac{\log^{[m-1]}T(r,f_u(g_v))}{\log^{[l-1]}T(r,g_v)} \le \min\left\{\frac{\overline{\sigma}^{(m,q,t)L}(f(g))}{\overline{\tau}^{(l,q,t)L}(g)}, \frac{\sigma^{(m,q,t)L}(f(g))}{\tau^{(l,q,t)L}(g)}\right\}$$
$$\le \max\left\{\frac{\overline{\sigma}^{(m,q,t)L}(f(g))}{\overline{\tau}^{(l,q,t)L}(g)}, \frac{\sigma^{(m,q,t)L}(f(g))}{\tau^{(l,q,t)L}(g)}\right\} \le \limsup_{r\to+\infty}\frac{\log^{[m-1]}T(r,f_u(g_v))}{\log^{[l-1]}T(r,g_v)}$$

for $m > 1$ *and* $l > 1$.

Theorem 4.3.22 *Let* $f(z)$ *be a meromorphic function and* $g(z)$ *be a non constant entire function such that* $0 < \overline{\tau}^{(m,q,t)L}(f(g)) \le \tau^{(m,q,t)L}(f(g)) < \infty$, $0 < \overline{\sigma}^{(l,q,t)L}(g) \le \sigma^{(l,q,t)L}(g) < \infty$, $\lambda^{(m,q,t)L}(f(g)) = \rho^{(l,q,t)L}(g)$. *Also let* f_u *and* g_v *be integer translations of* $f(z)$ *and* $g(z)$*, respectively, for* $u, v \in \mathbb{N}$. *If* $f_u(g_v) = h_t$*, where* h *is a meromorphic function and* $t \in \mathbb{N}$*, then*

$$\frac{t\cdot\ \overline{\tau}^{(1,q,t)L}(f(g))}{v\cdot\ \sigma^{(1,q,t)L}(g)} \le \liminf_{r\to+\infty}\frac{T(r,f_u(g_v))}{T(r,g_v)} \le \frac{t\cdot\ \overline{\tau}^{(1,q,t)L}(f(g))}{v\cdot\ \overline{\sigma}^{(1,q,t)L}(g)}$$
$$\le \limsup_{r\to+\infty}\frac{T(r,f_u(g_v))}{T(r,g_v)} \le \frac{t\cdot\ \tau^{(1,q,t)L}(f(g))}{v\cdot\ \overline{\sigma}^{(1,q,t)L}(g)}$$

and

$$\frac{\overline{\tau}^{(m,q,t)L}(f(g))}{\sigma^{(l,q,t)L}(g)} \leq \liminf_{r\to+\infty}\frac{\log^{[m-1]}T(r,f_u(g_v))}{\log^{[l-1]}T(r,g_v)} \leq \frac{\overline{\tau}^{(m,q,t)L}(f(g))}{\overline{\sigma}^{(l,q,t)L}(g)}$$
$$\leq \limsup_{r\to+\infty}\frac{\log^{[m-1]}T(r,f_u(g_v))}{\log^{[l-1]}T(r,g_v)} \leq \frac{\tau^{(m,q,t)L}(f(g))}{\overline{\sigma}^{(l,q,t)L}(g)}$$

for $m>1$ *and* $l>1$.

Theorem 4.3.23 *Let* $f(z)$ *be a meromorphic function and* $g(z)$ *be a non constant entire function such that* $0<\tau^{(m,q,t)L}(f(g))<\infty$, $0<\sigma^{(l,q,t)L}(g)<\infty$, $\lambda^{(m,q,t)L}(f(g))=\rho^{(l,q,t)L}(g)$. *Also let* f_u *and* g_v *be integer translations of* $f(z)$ *and* $g(z)$, *respectively, for* $u,v\in\mathbb{N}$. *If* $f_u(g_v)=h_t$, *where* h *is a meromorphic function and* $t\in\mathbb{N}$, *then*

$$\liminf_{r\to+\infty}\frac{T(r,f_u(g_v))}{T(r,g_v)} \leq \frac{t\cdot\tau^{(1,q,t)L}(f(g))}{v\cdot\sigma^{(1,q,t)L}(g)} \leq \limsup_{r\to+\infty}\frac{T(r,f_u(g_v))}{T(r,g_v)}$$

and

$$\liminf_{r\to+\infty}\frac{\log^{[m-1]}T(r,f_u(g_v))}{\log^{[l-1]}T(r,g_v)} \leq \frac{\tau^{(m,q,t)L}(f(g))}{\sigma^{(l,q,t)L}(g)} \leq \limsup_{r\to+\infty}\frac{\log^{[m-1]}T(r,f_u(g_v))}{\log^{[l-1]}T(r,g_v)}$$

for $m>1$ *and* $l>1$.

Theorem 4.3.24 *Let* $f(z)$ *be a meromorphic function and* $g(z)$ *be a non constant entire function such that* $0<\overline{\tau}^{(m,q,t)L}(f(g))\leq\tau^{(m,q,t)L}(f(g))<\infty$, $0<\overline{\sigma}^{(l,q,t)L}(g)\leq\sigma^{(l,q,t)L}(g)$ $<\infty$, $\lambda^{(m,q,t)L}(f(g))=\rho^{(l,q,t)L}(g)$. *Also let* f_u *and* g_v *be integer translations of* $f(z)$ *and* $g(z)$, *respectively, for* $u,v\in\mathbb{N}$. *If* $f_u(g_v)=h_t$, *where* h *is a meromorphic function and* $t\in\mathbb{N}$, *then*

$$\liminf_{r\to+\infty}\frac{T(r,f_u(g_v))}{T(r,g_v)} \leq \min\left\{\frac{t\cdot\overline{\tau}^{(1,q,t)L}(f(g))}{v\cdot\overline{\sigma}^{(1,q,t)L}(g)},\frac{t\cdot\tau^{(1,q,t)L}(f(g))}{v\cdot\sigma^{(1,q,t)L}(g)}\right\}$$
$$\leq \max\left\{\frac{t\cdot\overline{\tau}^{(1,q,t)L}(f(g))}{v\cdot\overline{\sigma}^{(1,q,t)L}(g)},\frac{t\cdot\tau^{(1,q,t)L}(f(g))}{v\cdot\sigma^{(1,q,t)L}(g)}\right\} \leq \limsup_{r\to+\infty}\frac{T(r,f_u(g_v))}{T(r,g_v)}$$

and

$$\liminf_{r\to+\infty}\frac{\log^{[m-1]}T(r,f_u(g_v))}{\log^{[l-1]}T(r,g_v)} \leq \min\left\{\frac{\overline{\tau}^{(m,q,t)L}(f(g))}{\overline{\sigma}^{(l,q,t)L}(g)},\frac{\tau^{(m,q,t)L}(f(g))}{\sigma^{(l,q,t)L}(g)}\right\}$$
$$\leq \max\left\{\frac{\overline{\tau}^{(m,q,t)L}(f(g))}{\overline{\sigma}^{(l,q,t)L}(g)},\frac{\tau^{(m,q,t)L}(f(g))}{\sigma^{(l,q,t)L}(g)}\right\} \leq \limsup_{r\to+\infty}\frac{\log^{[m-1]}T(r,f_u(g_v))}{\log^{[l-1]}T(r,g_v)}$$

for $m>1$ *and* $l>1$.

4.4 Concluding Remark

The results, as proved in this chapter, are no doubt powerful and illuminating in the context of the growth analysis of composite entire and meromorphic functions. But in order to have more powerful results, one may think about the concepts of functions of several complex variables in order to extend the earlier results, which are left to the interested readers.

References

[1] D. Somasundaram and R.Thamizharasi, "A note on the entire functions of L-bounded index and L-type", *Indian J. Pure Appl. Math.*, Vol. 19 , No.3, pp.284-293, March 1988.

[2] T. Biswas and S. K. Datta, "Effect of integer translation on relative order and relative type of entire and meromorphic functions", *Commun. Korean Math. Soc.*, Vol. 33, No. 2, pp. 485-494, 2018.

CHAPTER 5

$(p,q,t)L$-th Order and $(p,q,t)L$-th Type Based Some Growth Rates of Integer Translated Composite Entire and Meromorphic Functions

Abstract: In this chapter, we establish some new results depending on the comparative growth properties of the composition of integer translated entire and meromorphic functions using $(p,q,t)L$-th order and $(p,q,t)L$-th type.

Keywords: Growth rates, Integer translated entire function, Integer translated meromorphic function, $(p,q,t)L$-th order, $(p,q,t)L$-th type.
Mathematics Subject Classification (2020) : 30D30, 30D35.

5.1 Introduction

We denote by $\mathbb{C}$ the set of all finite complex numbers. Let $f(z)$ be an entire function defined on $\mathbb{C}$. The maximum modulus function corresponding to entire $f(z)$ is defined as $M(r,f) = \max\{|f(z)| : |z| = r\}$. When $f(z)$ is meromorphic, $M(r,f)$ can not be defined as $f(z)$ is not analytic. In this case, one may define another function $T(r,f)$, known as Nevanlinna's Characteristic function of $f(z)$, playing the same role as maximum modulus function in the following manner:

$$T(r,f) = N(r,f) + m(r,f),$$

where the function $N(r,f)$ and $m(r,f)$ are, respectively, the enumerative function and the proximity function corresponding to $f(z)$. For further details, one may see [1]. If $f(z)$ is an entire function, then Nevanlinna's Characteristic $T(r,f)$ of $f(z)$ reduces to $m(r,f)$.

Tanmay Biswas and Chinmay Biswas

Lakshminarasimhan [2] introduced the idea of the functions of the L-bounded index. Later Lahiri and Bhattacharjee [3] worked on the entire functions of the L-bounded index and of non uniform L-bounded index. In this chapter, we establish some new results depending on the comparative growth properties of the composition of integer translated entire and meromorphic functions using $(p,q,t)L$-th order and $(p,q,t)L$-th type. Indeed some works in this direction have also been explored in [4-6].

5.2 Lemma

In this section, we present some lemmas which will be needed in the sequel.

Lemma 5.2.1 [7] *If $f(z)$ is meromorphic and $g(z)$ is entire, then for all sufficiently large values of r,*

$$T(r, f(g)) \leq \{1+o(1)\} \frac{T(r,g)}{\log M(r,g)} T(M(r,g), f).$$

5.3 Main Results

In this section, we present the main results of the chapter.

Theorem 5.3.1 *Let $f(z)$ be a meromorphic function and $g(z)$ be a non constant entire function such that $\rho^{(m,n,t)L}(g) < \lambda^{(p,q,t)L}(f) \leq \rho^{(p,q,t)L}(f) < +\infty$ where $q \geq m$. Also let f_u and g_v be integer translations of $f(z)$ and $g(z)$, respectively, for $u, v \in \mathbb{N}$. Then*

$$\lim_{r\to+\infty} \frac{\log^{[p]} T(r, f_u(g_v))}{\log^{[p-m]} T(r, f_u)} = 0,$$

when for some $\alpha < \lambda^{(p,q,t)L}(f)$,
$\exp^{[t]} L(M(r,g)) = o\{\exp^{[m-1]}[(\log^{[q-1]} r)\exp^{[t+1]} L(r)]^{\alpha}\}$ *as $r \to +\infty$.*

Proof Let $f_u \circ g_v = h_t$, where is a meromorphic function and $t \in \mathbb{N}$. So h_t can be expressed as

$$h_t = \{(z+t) : t \in \mathbb{N}\}.$$

Then by (2.1.3) we obtain

$$T(r, h_t) = tT(r,h) + \sum_t (e_t + e_t^{'}),$$

where $e_t,\ e_t^{'} \to 0$ as $r \to +\infty$,

$$i.e., T(r, f_u \circ g_v) = tT(r, f \circ g) + \sum_t (e_t + e_t^{'}). \tag{5.3.1}$$

Now in view of Lemma 5.2.1, (5.3.1) and the inequality $T(r,g) \leq \log M(r,g)$ {cf. [5] } we get for all sufficiently large values of r that

$$\log^{[p]} T(r, f_u(g_v)) \leq$$
$$(\rho^{(p,q,t)L}(f) + \varepsilon)[\log^{[q]} M(r,g) + \exp^{[t]} L(M(r,g))] + O(1) \quad (5.3.2)$$
$$i.e., \; \log^{[p]} T(r, f_u(g_v)) \leq (\rho^{(p,q,t)L}(f) + \varepsilon)[\log^{[m]} M(r,g) + \exp^{[t]} L(M(r,g))] + O(1)$$
$$i.e., \; \log^{[p]} T(r, f_u(g_v)) \leq (\rho^{(p,q,t)L}(f) + \varepsilon) \times$$
$$[\exp^{[m-1]}[(\log^{[n-1]} r)\exp^{[t+1]} L(r)]^{(\rho^{(m,n,t)L}(g)+\varepsilon)} + \exp^{[t]} L(M(r,g))] + O(1). \quad (5.3.3)$$

Also in view of Lemma 2.2.2, we obtain for all sufficiently large values of r that

$$\log^{[p-m]} T(r, f_u) \geq \exp^{[m-1]}[(\log^{[q-1]} r)\exp^{[t+1]} L(r)]^{(\lambda^{(p,q,t)L}(f_u)-\varepsilon)}$$
$$i.e., \; \log^{[p-m]} T(r, f_u) \geq \exp^{[m-1]}[(\log^{[q-1]} r)\exp^{[t+1]} L(r)]^{(\lambda^{(p,q,t)L}(f)-\varepsilon)}. \quad (5.3.4)$$

Now from (5.3.3) and (5.3.4) we get for all sufficiently large values of r that

$$\frac{\log^{[p]} T(r, f_u(g_v))}{\log^{[p-m]} T(r, f_u)} \leq$$
$$\frac{(\rho^{(p,q,t)L}(f) + \varepsilon)[\exp^{[m-1]}[(\log^{[n-1]} r)\exp^{[t+1]} L(r)]^{(\rho^{(m,n,t)L}(g)+\varepsilon)}}{\exp^{[m-1]}[(\log^{[q-1]} r)\exp^{[t+1]} L(r)]^{(\lambda^{(p,q,t)L}(f)-\varepsilon)}}$$
$$+ \frac{\exp^{[t]} L(M(r,g))] + O(1)}{\exp^{[m-1]}[(\log^{[q-1]} r)\exp^{[t+1]} L(r)]^{(\lambda^{(p,q,t)L}(f)-\varepsilon)}}. \quad (5.3.5)$$

Since $\rho^{(m,n,t)L}(g) < \lambda^{(p,q,t)L}(f)$, we can choose $\varepsilon(>0)$ in such a way that

$$\rho^{(m,n,t)L}(g) + \varepsilon < \lambda^{(p,q,t)L}(f) - \varepsilon. \quad (5.3.6)$$

Now let for some $\alpha < \lambda^{(p,q,t)L}(f)$,
$\exp^{[t]} L(M(r,g)) = o\{\exp^{[m-1]}[(\log^{[q-1]} r)\exp^{[t+1]} L(r)]^{\alpha}\}$ as $r \to +\infty$.

As $\alpha < \lambda^{(p,q,t)L}(f)$ we can choose $\varepsilon(>0)$ in such a way that

$$\alpha < \lambda^{(p,q,t)L}(f) - \varepsilon. \quad (5.3.7)$$

Since $\exp^{[t]} L(M(r,g)) = o\{\exp^{[m-1]}[(\log^{[q-1]} r)\exp^{[t+1]} L(r)]^{\alpha}\}$ as $r \to +\infty$ we get on using (5.3.7) that

$$\frac{\exp^{[t]} L(M(r,g))}{\exp^{[m-1]}[(\log^{[q-1]} r)\exp^{[t+1]} L(r)]^{\alpha}} \to 0 \text{ as } r \to +\infty$$
$$i.e., \; \frac{\exp^{[t]} L(M(r,g))}{\exp^{[m-1]}[(\log^{[q-1]} r)\exp^{[t+1]} L(r)]^{(\lambda^{(p,q,t)L}(f)-\varepsilon)}} \to 0 \text{ as } r \to +\infty. \quad (5.3.8)$$

Now in view of (5.3.5), (5.3.6) and (5.3.8) we obtain that

$$\lim_{r \to +\infty} \frac{\log^{[p]} T(r, f_u(g_v))}{\log^{[p-m]} T(r, f_u)} = 0.$$

Thus the theorem follows.

Theorem 5.3.2 *Let $f(z)$ be a meromorphic function and $g(z)$ be a non constant entire function such that $\rho^{(p,q,t)L}(f) < +\infty$, $\lambda^{(m,n,t)L}(g) > 0$ and $\rho^{(m,n,t)L}(g) < +\infty$ where $m > q$. Also let f_u and g_v be integer translations of $f(z)$ and $g(z)$, respectively, for $u, v \in \mathbb{N}$. Then*

$$\limsup_{r\to+\infty} \frac{\log^{[p+m-q]} T(r, f_u(g_v))}{\log^{[m]} T(r, g_v) + \exp^{[t]} L(M(r,g))} \leq \frac{\rho^{(m,n,t)L}(g)}{\lambda^{(m,n,t)L}(g)},$$

when $\exp^{[t]} L(M(r,g)) = o\{\log^{[m]} T(r, g_v)\}$ *as* $r \to +\infty$.

Proof From (5.3.2) for all sufficiently large values of r,we have

$$\log^{[p]} T(r, f_u(g_v)) \leq (\rho^{(p,q,t)L}(f) + \varepsilon)[\log^{[q]} M(r,g) + \exp^{[t]} L(M(r,g)) + O(1)]$$

$$i.e., \log^{[p]} T(r, f_u(g_v)) \leq (\rho^{(p,q,t)L}(f) + \varepsilon) \cdot \log^{[q]} M(r,g) + (\rho^{(p,q,t)L}(f) + \varepsilon) \cdot [\exp^{[t]} L(M(r,g)) + O(1)]$$

$$i.e., \log^{[p]} T(r, f_u(g_v)) \leq (\rho^{(p,q,t)L}(f) + \varepsilon) \cdot \log^{[q]} M(r,g) \left[\frac{(\rho^{(p,q,t)L}(f) + \varepsilon) \cdot \log^{[q]} M(r,g)}{(\rho^{(p,q,t)L}(f) + \varepsilon) \cdot \log^{[q]} M(r,g)} + \frac{(\rho^{(p,q,t)L}(f) + \varepsilon) \cdot [\exp^{[t]} L(M(r,g)) + O(1)}{(\rho^{(p,q,t)L}(f) + \varepsilon) \cdot \log^{[q]} M(r,g)}\right]$$

$$i.e., \log^{[p]} T(r, f_u(g_v)) \leq (\rho^{(p,q,t)L}(f) + \varepsilon) \cdot \log^{[q]} M(r,g) \left[1 + \frac{\exp^{[t]} L(M(r,g)) + O(1)}{\log^{[q]} M(r,g)}\right]$$

$$i.e., \log^{[p+1]} T(r, f_u(g_v)) \leq \log(\rho^{(p,q,t)L}(f) + \varepsilon) + \log^{[q+1]} M(r,g) + \log\left[1 + \frac{\exp^{[t]} L(M(r,g)) + O(1)}{\log^{[q]} M(r,g)}\right]$$

Taking $\log\left(1 + \frac{\exp^{[t]} L(M(r,g)) + O(1)}{\log^{[q]} M(r,g)}\right) \sim \frac{\exp^{[t]} L(M(r,g)) + O(1)}{\log^{[q]} M(r,g)}$, we get for all sufficiently large values of r

$$\log^{[p+1]} T(r, f_u(g_v)) \leq \log^{[q+1]} M(r,g) + \log(\rho^{(p,q,t)L}(f) + \varepsilon) + \frac{\exp^{[t]} L(M(r,g)) + O(1)}{\log^{[q]} M(r,g)}$$

$$i.e., \quad \log^{[p+1]} T(r, f_u(g_v)) \leq \log^{[q+1]} M(r,g) \left[1 + \frac{\exp^{[t]} L(M(r,g)) + O(1) + \log^{[q]} M(r,g) \cdot \log(\rho^{(p,q,t)L}(f) + \varepsilon)}{\log^{[q]} M(r,g) \cdot \log^{[q+1]} M(r,g)}\right].$$

$$i.e., \ \log^{[p+2]} T(r, f_u(g_v)) \leq \log^{[q+2]} M(r,g)$$
$$\log\left[1+\frac{\exp^{[t]} L(M(r,g))+O(1)+\log^{[q]} M(r,g)\cdot\log(\rho^{(p,q,t)L}(f)+\varepsilon)}{\log^{[q]} M(r,g)\cdot\log^{[q+1]} M(r,g)}\right].$$

Again using $\log(1+x)\sim x$ for $x=\frac{\exp^{[t]} L(M(r,g))+O(1)+\log^{[q]} M(r,g)\cdot\log(\rho^{(p,q,t)L}(f)+\varepsilon)}{\prod_{k=q}^{q+1}\log^{[k]} M(r,g)}$, we get from above for all sufficiently large positive numbers of r ,

$$\log^{[p+2]} T(r, f_u(g_v)) \leq \log^{[q+2]} M(r,g) + \frac{\exp^{[t]} L(M(r,g))+O(1)+\log^{[q]} M(r,g)\cdot\log(\rho^{(p,q,t)L}(f)+\varepsilon)}{\prod_{k=q}^{q+1}\log^{[k]} M(r,g)}.$$

Continuing this process, we get

$$\log^{[p+m-q]} T(r, f_u(g_v)) \leq \log^{[q+m-q]} M(r,g) + \frac{\exp^{[t]} L(M(r,g))+O(1)+\log^{[q]} M(r,g)\cdot\log(\rho^{(p,q,t)L}(f)+\varepsilon)}{\prod_{k=q}^{q+m-q-1}\log^{[k]} M(r,g)}.$$

$$i.e., \log^{[p+m-q]} T(r, f_u(g_v)) \leq \log^{[m]} M(r,g) + \frac{\exp^{[t]} L(M(r,g))+O(1)+\log^{[q]} M(r,g)\cdot\log(\rho^{(p,q,t)L}(f)+\varepsilon)}{\prod_{k=q}^{m-1}\log^{[k]} M(r,g)}.$$

$$i.e., \ \log^{[p+m-q]} T(r, f_u(g_v)) \leq (\rho^{(m,n,t)L}(g)+\varepsilon)[\log^{[n]} r+\exp^{[t]} L(r)] + \frac{\exp^{[t]} L(M(r,g))+O(1)+\log^{[q]} M(r,g)\cdot\log(\rho^{(p,q,t)L}(f)+\varepsilon)}{\prod_{k=q}^{m-1}\log^{[k]} M(r,g)}. \tag{5.3.9}$$

Again in view of Lemma 2.2.2, we have for all sufficiently large values of r that

$$\log^{[m]} T(r, g_v) \geq (\lambda^{(m,n,t)L}(g_v)-\varepsilon)[\log^{[n]} r+\exp^{[t]} L(r)]$$
$$i.e., \ \log^{[n]} r+\exp^{[t]} L(r) \leq \frac{\log^{[m]} T(r, g_v)}{(\lambda^{(m,n,t)L}(g)-\varepsilon)}. \tag{5.3.10}$$

Hence from (5.3.9) and (5.3.10), it follows for all sufficiently large values of r that

$$\log^{[p+m-q]} T(r, f_u(g_v)) \leq \left(\frac{\rho^{(m,n,t)L}(g)+\varepsilon}{\lambda^{(m,n,t)L}(g_v)-\varepsilon}\right)\cdot\log^{[m]} T(r, g_v)$$

$$+\frac{\exp^{[t]}L(M(r,g))+O(1)+\log^{[q]}M(r,g)\cdot\log(\rho^{(p,q,t)L}(f)+\varepsilon)}{\prod_{k=q}^{m-1}\log^{[k]}M(r,g)}$$

$$i.e,\ \frac{\log^{[p+m-q]}T(r,f_u(g_v))}{\log^{[m]}T(r,g_v)+\exp^{[t]}L(M(r,g))}$$

$$\leq\left(\frac{\rho^{(m,n,t)L}(g)+\varepsilon}{\lambda^{(m,n,t)L}(g)-\varepsilon}\right)\cdot\frac{\log^{[m]}T(r,g_v)}{\log^{[m]}T(r,g_v)+\exp^{[t]}L(M(r,g))}$$

$$+\frac{\exp^{[t]}L(M(r,g))+O(1)+\log^{[q]}M(r,g)\cdot\log(\rho^{(p,q,t)L}(f)+\varepsilon)}{[\log^{[m]}T(r,g_v)+\exp^{[t]}L(M(r,g))]\cdot\prod_{k=q}^{m-1}\log^{[k]}M(r,g)}$$

$$i.e,\ \frac{\log^{[p+m-q]}T(r,f_u(g_v))}{\log^{[m]}T(r,g_v)+\exp^{[t]}L(M(r,g))}\leq$$

$$\frac{\frac{\rho^{(m,n,t)L}(g)+\varepsilon}{\lambda^{(m,n,t)L}(g)-\varepsilon}}{1+\frac{\exp^{[t]}L(M(r,g))}{\log^{[m]}T(r,g_v)}}+\frac{1+\frac{O(1)+\log^{[q]}M(r,g)\cdot\log(\rho^{(p,q,t)L}(f)+\varepsilon)}{\exp^{[t]}L(M(r,g))}}{[1+\frac{\log^{[m]}T(r,g_v)}{\exp^{[t]}L(M(r,g))}]\cdot\prod_{k=q}^{m-1}\log^{[k]}M(r,g)}.$$

$$i.e,\ \frac{\log^{[p+m-q]}T(r,f_u(g_v))}{\log^{[m]}T(r,g_v)+\exp^{[t]}L(M(r,g))}\leq\frac{\frac{\rho^{(m,n,t)L}(g)+\varepsilon}{\lambda^{(m,n,t)L}(g)-\varepsilon}}{1+\frac{\exp^{[t]}L(M(r,g))}{\log^{[m]}T(r,g_v)}}$$

$$+\frac{\frac{1}{\prod_{k=q}^{m-1}\log^{[k]}M(r,g)}+\frac{O(1)}{\exp^{[t]}L(M(r,g))\cdot\prod_{k=q}^{m-1}\log^{[k]}M(r,g)}+\frac{\log(\rho^{(p,q,t)L}(f)+\varepsilon)}{\exp^{[t]}L(M(r,g))\cdot\prod_{k=q+1}^{m-1}\log^{[k]}M(r,g)}}{[1+\frac{\log^{[m]}T(r,g_v)}{\exp^{[t]}L(M(r,g))}]}.$$

(5.3.11)

Since $\exp^{[t]}L(M(r,g))=o\{\log^{[m]}T(r,g_v)\}$ as $r\to+\infty$ and $\varepsilon(>0)$ is arbitrary we obtain from (5.3.11) that

$$\limsup_{r\to+\infty}\frac{\log^{[p+m-q]}T(r,f_u(g_v))}{\log^{[m]}T(r,g_v)+\exp^{[t]}L(M(r,g))}\leq\frac{\rho^{(m,n,t)L}(g)}{\lambda^{(m,n,t)L}(g)}.$$

Thus the theorem is established.

In the line of Theorem 5.3.2, the following theorem may be proved and therefore its proof is omitted:

Theorem 5.3.3 *Let $f(z)$ be a meromorphic function and $g(z)$ be a non constant entire function such that $\rho^{(p,q,t)L}(f)<+\infty$, $\lambda^{(m,n,t)L}(g)>0$ and either $0<\lambda^{(m,n,t)L}(g)<+\infty$ or $0<\rho^{(m,n,t)L}(g)<+\infty$ hold where $m>q$. Also let f_u and g_v be integer translations of $f(z)$ and $g(z)$, respectively, for $u,v\in\mathbb{N}$. Then*

$$\liminf_{r\to+\infty}\frac{\log^{[p+m-q]}T(r,f_u(g_v))}{\log^{[m]}T(r,g_v)+\exp^{[t]}L(M(r,g))}\leq 1,$$

when $\exp^{[t]} L(M(r,g)) = o\{\log^{[m]} T(r, g_v)\}$ *as* $r \to +\infty$.

Now we state the following three theorems without their proofs as those can be carried out in the line of Theorem 5.3.2 and Theorem 5.3.3

Theorem 5.3.4 *Let* $f(z)$ *be a meromorphic function and* $g(z)$ *be a non constant entire function such that* $0 < \lambda^{(p,q,t)L}(f) \leq \rho^{(p,q,t)L}(f) < +\infty$ *and* $\rho^{(m,n,t)L}(g) < +\infty$ *where* $m > n = q$. *Also let* f_u *and* g_v *be integer translations of* $f(z)$ *and* $g(z)$, *respectively, for* $u, v \in \mathbb{N}$. *Then*

$$\limsup_{r \to +\infty} \frac{\log^{[p+m-q]} T(r, f_u(g_v))}{\log^{[p]} T(r, f_u) + \exp^{[t]} L(M(r,g))} \leq \frac{\rho^{(m,n,t)L}(g)}{\lambda^{(p,q,t)L}(f)},$$

when $\exp^{[t]} L(M(r,g)) = o\{\log^{[p]} T(r, f_u)\}$ *as* $r \to +\infty$.

Theorem 5.3.5 *Let* $f(z)$ *be a meromorphic function and* $g(z)$ *be a non constant entire function such that* $0 < \rho^{(p,q,t)L}(f) < +\infty$ *and* $\rho^{(m,n,t)L}(g) < +\infty$ *where* $m > n = q$. *Also let* f_u *and* g_v *be integer translations of* $f(z)$ *and* $g(z)$, *respectively, for* $u, v \in \mathbb{N}$. *Then*

$$\liminf_{r \to +\infty} \frac{\log^{[p+m-q]} T(r, f_u(g_v))}{\log^{[p]} T(r, f_u) + \exp^{[t]} L(M(r,g))} \leq \frac{\rho^{(m,n,t)L}(g)}{\rho^{(p,q,t)L}(f)},$$

when $\exp^{[t]} L(M(r,g)) = o\{\log^{[p]} T(r, f_u)\}$ *as* $r \to +\infty$.

Theorem 5.3.6 *Let* $f(z)$ *be a meromorphic function and* $g(z)$ *be a non constant entire function such that* $0 < \lambda^{(p,q,t)L}(f) \leq \rho^{(p,q,t)L}(f) < +\infty$ *and* $\lambda^{(m,n,t)L}(g) < +\infty$ *where* $m > n = q$. *Also let* f_u *and* g_v *be integer translations of* $f(z)$ *and* $g(z)$, *respectively, for* $u, v \in \mathbb{N}$. *Then*

$$\liminf_{r \to +\infty} \frac{\log^{[p+m-q]} T(r, f_u(g_v))}{\log^{[p]} T(r, f_u) + \exp^{[t]} L(M(r,g))} \leq \frac{\rho^{(m,n,t)L}(g)}{\lambda^{(p,q,t)L}(f)},$$

when $\exp^{[t]} L(M(r,g)) = o\{\log^{[p]} T(r, f_u)\}$ *as* $r \to +\infty$.

Theorem 5.3.7 *Let* $f(z)$ *be a meromorphic function and* $g(z)$ *be a non constant entire function such that* $\rho^{(m,q,t)L}(f) < \infty$ *and* $\lambda^{(p,q,t)L}(f(g)) = +\infty$. *Also let* f_u *and* g_v *be integer translations of* $f(z)$ *and* $g(z)$, *respectively, for* $u, v \in \mathbb{N}$. *Then*

$$\lim_{r \to +\infty} \frac{\log^{[p]} T(r, f_u(g_v))}{\log^{[m]} T(r, f_u)} = +\infty.$$

Proof If possible, let there exists a constant β such that for a sequence of values of r tending to infinity we have

$$\log^{[p]} T(r, f_u(g_v)) \leq \beta \cdot \log^{[m]} T(r, f_u). \tag{5.3.12}$$

Again from the definition of $\rho^{(m,q,t)L}(f_u)$, it follows in view of Lemma 2.2.2 for all sufficiently large values of r that

$$\log^{[m]} T(r, f_u) \leq (\rho^{(m,q,t)L}(f_u) + \varepsilon)[\log^{[q]} r + \exp^{[t]} L(r)]$$

$$i.e., \ \log^{[m]} T(r, f_u) \leq (\rho^{(m,q,t)L}(f) + \varepsilon)[\log^{[q]} r + \exp^{[t]} L(r)]. \tag{5.3.13}$$

Now combining (5.3.12) and (5.3.13) we obtain for a sequence of values of r tending to infinity that

$$\log^{[p]} T(r, f_u(g_v)) \leq \beta \cdot (\rho^{(m,q,t)L}(f) + \varepsilon)[\log^{[q]} r + \exp^{[t]} L(r)]$$
$$i.e., \ \lambda^{(p,q,t)L}(f_u(g_v)) = \lambda^{(p,q,t)L}(f(g)) \leq \beta \cdot (\rho^{(m,q,t)L}(f) + \varepsilon),$$

which contradicts the condition $\lambda^{(p,q,t)L}(f \circ g) = \infty$. Hence the theorem follows.

Remark 5.3.1 *Theorem 5.3.7 is also valid with "limit superior" instead of "limit" if $\lambda^{(p,q,t)L}(f(g)) = +\infty$ is replaced by $\rho^{(p,q,t)L}(f(g)) = +\infty$ and the other conditions remaining the same.*

Corollary 5.3.1 *Under the assumptions of Theorem 5.3.7 and Remark 5.3.1,*

$$\lim_{r\to+\infty} \frac{\log^{[p-1]} T(r, f_u(g_v))}{\log^{[m-1]} T(r, f_u)} = +\infty \quad and \quad \limsup_{r\to+\infty} \frac{\log^{[p-1]} T(r, f_u(g_v))}{\log^{[m-1]} T(r, f_u)} = +\infty$$

respectively holds.

Proof From Theorem 5.3.7 we obtain for all sufficiently large positive numbers of r and for $K > 1$,

$$\log^{[p]} T(r, f_u(g_v)) > K \log^{[m]} T(r, f_u)$$
$$i.e., \ \log^{[p-1]} T(r, f_u(g_v)) > \left\{\log^{[m-1]} T(r, f_u)\right\}^K,$$

from which the first part of the corollary follows.

Similarly, using Remark 5.3.1, we obtain the second part of the corollary.

In the line of Theorem 5.3.7, one can easily prove the following theorem and therefore its proof is omitted.

Theorem 5.3.8 *Let $f(z)$ be a meromorphic function and $g(z)$ be a non constant entire function such that $\rho^{(m,q,t)L}(g) < +\infty$ and $\lambda^{(p,q,t)L}(f(g)) = +\infty$. Also let f_u and g_v be integer translations of $f(z)$ and $g(z)$, respectively, for $u, v \in \mathbb{N}$. Then*

$$\lim_{r\to+\infty} \frac{\log^{[p]} T(r, f_u(g_v))}{\log^{[m]} T(r, g_v)} = +\infty.$$

We omit the proof of Theorem 5.3.8 because it can be carried out in the line of Theorem 5.3.7.

Remark 5.3.2 *Theorem 5.3.8 is also valid with "limit superior" instead of "limit" if $\lambda^{(p,q,t)L}(f(g)) = +\infty$ is replaced by $\rho^{(p,q,t)L}(f(g)) = +\infty$ and the other conditions remaining the same.*

In the line of Corollary 5.3.1, one can easily verify the following corollary:

Corollary 5.3.2 *Under the assumptions of Theorem 5.3.8 and Remark 5.3.2,*

$$\lim_{r\to+\infty} \frac{\log^{[p-1]} T(r, f_u(g_v))}{\log^{[m-1]} T(r, g_v)} = +\infty \quad \text{and} \quad \limsup_{r\to+\infty} \frac{\log^{[p-1]} T(r, f_u(g_v))}{\log^{[m-1]} T(r, g_v)} = +\infty$$

respectively hold.

Theorem 5.3.9 *Let $f(z)$ be a meromorphic function and $g(z)$ be a non constant entire function such that $0 < \lambda^{(p,q,t)L}(f) \leq \rho^{(p,q,t)L}(f) < +\infty$ and $\sigma^{(m,n,t)L}(g) < +\infty$ where $m-1 \leq q$, $a > 1$. Also let f_u and g_v be integer translations of $f(z)$ and $g(z)$, respectively, for $u, v \in \mathbb{N}$. Then*

$$\limsup_{r\to+\infty} \frac{\log^{[p]} T(r, f_u(g_v))}{\log^{[p]} T(\exp^{[q]}[\log^{[n-1]} r \cdot \exp^{[t+1]} L(r)]^{\rho^{(m,n,t)L}(g)}, f_u)}$$
$$\leq \frac{\rho^{(p,q,t)L}(f) \cdot \sigma^{(m,n,t)L}(g)}{\lambda^{(p,q,t)L}(f)},$$

when for some positive $\alpha < \rho^{(m,n,t)L}(g)$,
$\exp^{[t]} L(M(r,g)) = o([\log^{[n-1]} r \cdot \exp^{[t+1]} L(r)]^{\alpha})$ *as* $r \to +\infty$.

Proof Since $0 < \rho^{(p,q,t)L}(f) < \infty$ and $L(r)$ is an increasing function of r, it follows from (5.3.2) for all sufficiently large values of r that

$$\log^{[p]} T(r, f_u(g_v)) \leq (\rho^{(p,q,t)L}(f) + \varepsilon)[\log^{[q]} M(r,g) + \exp^{[t]} L(M(r,g))] + O(1)$$

$$i.e.,\ \log^{[p]} T(r, f_u(g_v)) \leq (\rho^{(p,q,t)L}(f) + \varepsilon)[\log^{[m-1]} M(r,g) + \exp^{[t]} L(M(r,g))] + O(1)$$

$$i.e.,\ \log^{[p]} T(r, f_u(g_v)) \leq (\rho^{(p,q,t)L}(f) + \varepsilon) \cdot$$
$$[(\sigma^{(m,n,t)L}(g) + \varepsilon)[\log^{[n-1]} r \cdot \exp^{[t+1]} L(r)]^{\rho^{(m,n,t)L}(g)} + \exp^{[t]} L(M(r,g))] + O(1). \quad (5.3.14)$$

Also, we obtain in view of Lemma 2.2.2 for all sufficiently large values of r that

$$\log^{[p]} T(\exp^{[q]}[\log^{[n-1]} r \cdot \exp^{[t+1]} L(r)]^{\rho^{(m,n,t)L}(g)}, f_u) \geq$$
$$(\lambda^{(p,q,t)L}(f_u) - \varepsilon)[\log^{[n-1]} r \cdot \exp^{[t+1]} L(r)]^{\rho^{(m,n,t)L}(g)}$$
$$+ (\lambda^{(p,q,t)L}(f_u) - \varepsilon) \exp^{[t]}[L(\exp^{[q]}[\log^{[n-1]} r \cdot \exp^{[t+1]} L(r)]^{\rho^{(m,n,t)L}(g)})]$$

$$i.e.,\ \log^{[p]} T(\exp^{[q]}[\log^{[n-1]} r \cdot \exp^{[t+1]} L(r)]^{\rho^{(m,n,t)L}(g)}, f_u) \geq$$
$$(\lambda^{(p,q,t)L}(f) - \varepsilon)[\log^{[n-1]} r \cdot \exp^{[t+1]} L(r)]^{\rho^{(m,n,t)L}(g)}$$

$$+ (\lambda^{(p,q,t)L}(f) - \varepsilon) \exp^{[t]}[L(\exp^{[q]}[\log^{[n-1]} r \cdot \exp^{[t+1]} L(r)]^{\rho^{(m,n,t)L}(g)})]$$

$$\log^{[p]} T(\exp^{[q]}[\log^{[n-1]} r \cdot \exp^{[t+1]} L(r)]^{\rho^{(m,n,t)L}(g)}, f_u) \geq (\lambda^{(p,q,t)L}(f) - \varepsilon)[\log^{[n-1]} r \cdot \exp^{[t+1]} L(r)]^{\rho^{(m,n,t)L}(g)}.$$

Now from (5.3.14) and above it follows for all sufficiently large values of r that

$$\frac{\log^{[p]} T(r, f_u(g_v))}{\log^{[p]} T(\exp^{[q]}[\log^{[n-1]} r \cdot \exp^{[t+1]} L(r)]^{\rho^{(m,n,t)L}(g)}, f_u)}$$
$$\leq \frac{(\rho^{(p,q,t)L}(f) + \varepsilon)[(\sigma^{(m,n,t)L}(g) + \varepsilon)[\log^{[n-1]} r \cdot \exp^{[t+1]} L(r)]^{\rho^{(m,n,t)L}(g)}}{(\lambda^{(p,q,t)L}(f) - \varepsilon)[\log^{[n-1]} r \cdot \exp^{[t+1]} L(r)]^{\rho^{(m,n,t)L}(g)}}$$
$$+ \frac{\exp^{[t]} L(M(r, g))]}{(\lambda^{(p,q,t)L}(f) - \varepsilon)[\log^{[n-1]} r \cdot \exp^{[t+1]} L(r)]^{\rho^{(m,n,t)L}(g)}}$$
$$+ \frac{O(1)}{(\lambda^{(p,q,t)L}(f) - \varepsilon)[\log^{[n-1]} r \cdot \exp^{[t+1]} L(r)]^{\rho^{(m,n,t)L}(g)}} \tag{5.3.15}$$

As $\alpha < \rho^{(m,n,t)L}(g)$ and $\exp^{[t]} L(M(r, g)) = o([\log^{[n-1]} r \cdot \exp^{[t+1]} L(r)]^{\alpha})$ as $r \to +\infty$, we obtain that

$$\lim_{r \to +\infty} \frac{\exp^{[t]} L(M(r, g))}{[\log^{[n-1]} r \cdot \exp^{[t+1]} L(r)]^{\rho^{(m,n,t)L}(g)}} = 0. \tag{5.3.16}$$

Since $\varepsilon(> 0)$ is arbitrary, it follows from (5.3.15) and (5.3.16) that

$$\limsup_{r \to +\infty} \frac{\log^{[p]} T(r, f_u(g_v))}{\log^{[p]} T(\exp^{[q]}[\log^{[n-1]} r \cdot \exp^{[t+1]} L(r)]^{\rho^{(m,n,t)L}(g)}, f_u)} \leq \frac{\rho^{(p,q,t)L}(f) \cdot \sigma^{(m,n,t)L}(g)}{\lambda^{(p,q,t)L}(f)}.$$

In the line of Theorem 5.3.9, one can easily prove the following theorems and therefore their proofs are omitted.

Theorem 5.3.10 *Let $f(z)$ be a meromorphic function and $g(z)$ be a non constant entire function such that $0 < \lambda^{(p,q,t)L}(f) < +\infty$ or $0 < \rho^{(p,q,t)L}(f) < +\infty$ and $\sigma^{(m,n,t)L}(g) < +\infty$ where $m - 1 \leq q$, $a > 1$. Also let f_u and g_v be integer translations of $f(z)$ and $g(z)$, respectively, for $u, v \in \mathbb{N}$. Then*

$$\liminf_{r \to +\infty} \frac{\log^{[p]} T(r, f_u(g_v))}{\log^{[p]} T(\exp^{[q]}[\log^{[n-1]} r \cdot \exp^{[t+1]} L(r)]^{\rho^{(m,n,t)L}(g)}, f_u)} \leq \sigma^{(m,n,t)L}(g),$$

when $\exp^{[t]} L(M(r, g)) = o([\log^{[n-1]} r \cdot \exp^{[t+1]} L(r)]^{\alpha})$ as $r \to +\infty$ and for some positive $\alpha < \rho^{(m,n,t)L}(g)$.

Theorem 5.3.11 *Let $f(z)$ be a meromorphic function and $g(z)$ be a non constant entire function such that $0 < \lambda^{(p,q,t)L}(f) < +\infty$ or $0 < \rho^{(p,q,t)L}(f) < +\infty$ and $\tau^{(m,n,t)L}(g) < +\infty$ where $m-1 \leq q$, $a > 1$. Also let f_u and g_v be integer translations of $f(z)$ and $g(z)$, respectively, for $u, v \in \mathbb{N}$. Then*

$$\liminf_{r\to+\infty} \frac{\log^{[p]} T_h^{-1}(T_{f\circ g}(r))}{\log^{[p]} T(\exp^{[q]}[\log^{[n-1]} r \cdot \exp^{[t+1]} L(r)]^{\lambda^{(m,n,t)L}(g)}, f_u)} \leq \tau^{(m,n,t)L}(g),$$

when $\exp^{[t]} L(M(r,g)) = o([\log^{[n-1]} r \cdot \exp^{[t+1]} L(r)]^{\alpha})$ as $r \to +\infty$ and for some positive $\alpha < \lambda^{(m,n,t)L}(g)$.

Theorem 5.3.12 *Let $f(z)$ be a meromorphic function and $g(z)$ be a non constant entire function such that $0 < \lambda^{(p,q,t)L}(f) \leq \rho^{(p,q,t)L}(f) < +\infty$ and $\overline{\sigma}^{(m,n,t)L}(g) < +\infty$ where $m-1 \leq q$, $a > 1$. Also let f_u and g_v be integer translations of $f(z)$ and $g(z)$, respectively, for $u, v \in \mathbb{N}$. Then*

$$\liminf_{r\to+\infty} \frac{\log^{[p]} T_h^{-1}(T_{f\circ g}(r))}{\log^{[p]} T(\exp^{[q]}[\log^{[n-1]} r \cdot \exp^{[t+1]} L(r)]^{\rho^{(m,n,t)L}(g)}, f_u)} \leq \frac{\rho_h^{(p,q,t)L}(f) \cdot \overline{\sigma}^{(m,n,t)L}(g)}{\lambda_h^{(p,q,t)L}(f)},$$

when $\exp^{[t]} L(M(r,g)) = o([\log^{[n-1]} r \cdot \exp^{[t+1]} L(r)]^{\alpha})$ as $r \to +\infty$ and for some positive $\alpha < \rho^{(m,n,t)L}(g)$.

Theorem 5.3.13 *Let $f(z)$ be a meromorphic function and $g(z)$ be a non constant entire function such that $0 < \lambda^{(p,q,t)L}(f) \leq \rho^{(p,q,t)L}(f) < +\infty$ and $\overline{\tau}^{(m,n,t)L}(g) < +\infty$ where $m-1 \leq q$, $a > 1$. Also let f_u and g_v be integer translations of $f(z)$ and $g(z)$, respectively, for $u, v \in \mathbb{N}$. Then*

$$\liminf_{r\to+\infty} \frac{\log^{[p]} T_h^{-1}(T_{f\circ g}(r))}{\log^{[p]} T(\exp^{[q]}[\log^{[n-1]} r \cdot \exp^{[t+1]} L(r)]^{\lambda^{(m,n,t)L}(g)}, f_u)} \leq \frac{\rho_h^{(p,q,t)L}(f) \cdot \overline{\tau}^{(m,n,t)L}(g)}{\lambda_h^{(p,q,t)L}(f)},$$

when $\exp^{[t]} L(M(r,g)) = o([\log^{[n-1]} r \cdot \exp^{[t+1]} L(r)]^{\alpha})$ as $r \to +\infty$ and for some positive $\alpha < \lambda^{(m,n,t)L}(g)$.

Now we state the following theorems without their proofs as those can be carried out in the line of Theorem 5.3.9 .

Theorem 5.3.14 *Let $f(z)$ be a meromorphic function and $g(z)$ be a non constant entire function such that $\lambda^{(m,n,t)L}(g) > 0$, $\rho^{(p,q,t)L}(f) < +\infty$ and $\sigma^{(m,n,t)L}(g) < +\infty$ where $m-1 \leq q$. Also let f_u and g_v be integer translations of $f(z)$ and $g(z)$, respectively, for $u, v \in \mathbb{N}$. Then*

$$\limsup_{r\to+\infty} \frac{\log^{[p]} T(r, f_u(g_v))}{\log^{[m]} T(\exp^{[n]}[\log^{[n-1]} r \cdot \exp^{[t+1]} L(r)]^{\rho^{(m,n,t)L}(g)}), g_v)} \leq \frac{\rho^{(p,q,t)L}(f) \cdot \sigma^{(m,n,t)L}(g)}{\lambda_k^{(m,n,t)L}(g)},$$

when $\exp^{[t]} L(M(r,g)) = o([\log^{[n-1]} r \cdot \exp^{[t+1]} L(r)]^{\alpha})$ as $r \to +\infty$ and for some positive $\alpha < \rho^{(m,n,t)L}(g)$.

Theorem 5.3.15 *Let $f(z)$ be a meromorphic function and $g(z)$ be a non constant entire function such that $\lambda^{(m,n,t)L}(g) > 0$, $\lambda^{(p,q,t)L}(f) < +\infty$ and $\sigma^{(m,n,t)L}(g) < +\infty$ where $m-1 \leq q$. Also let f_u and g_v be integer translations of $f(z)$ and $g(z)$, respectively, for $u, v \in \mathbb{N}$. Then*

$$\liminf_{r\to+\infty} \frac{\log^{[p]} T(r, f_u(g_v))}{\log^{[m]} T(\exp^{[n]}[\log^{[n-1]} r \cdot \exp^{[t+1]} L(r)]^{\rho^{(m,n,t)L}(g)}), g_v)} \leq \frac{\lambda^{(p,q,t)L}(f) \cdot \sigma^{(m,n,t)L}(g)}{\lambda^{(m,n,t)L}(g)},$$

when $\exp^{[t]} L(M(r,g)) = o([\log^{[n-1]} r \cdot \exp^{[t+1]} L(r)]^{\alpha})$ as $r \to +\infty$ and for some positive $\alpha < \rho^{(m,n,t)L}(g)$.

Theorem 5.3.16 *Let $f(z)$ be a meromorphic function and $g(z)$ be a non constant entire function such that $\rho^{(m,n,t)L}(g) > 0$, $\rho^{(p,q,t)L}(f) < +\infty$ and $\sigma^{(m,n,t)L}(g) < +\infty$ where $m-1 \leq q$. Also let f_u and g_v be integer translations of $f(z)$ and $g(z)$, respectively, for $u, v \in \mathbb{N}$. Then*

$$\liminf_{r\to+\infty} \frac{\log^{[p]} T(r, f_u(g_v))}{\log^{[m]} T(\exp^{[n]}[\log^{[n-1]} r \cdot \exp^{[t+1]} L(r)]^{\rho^{(m,n,t)L}(g)}), g_v)} \leq \frac{\rho^{(p,q,t)L}(f) \cdot \sigma^{(m,n,t)L}(g)}{\rho^{(m,n,t)L}(g)},$$

when $\exp^{[t]} L(M(r,g)) = o([\log^{[n-1]} r \cdot \exp^{[t+1]} L(r)]^{\alpha})$ as $r \to +\infty$ and for some positive $\alpha < \rho^{(m,n,t)L}(g)$.

Theorem 5.3.17 *Let $f(z)$ be a meromorphic function and $g(z)$ be a non constant entire function such that $\lambda^{(m,n,t)L}(g) > 0$, $\rho^{(p,q,t)L}(f) < +\infty$ and $\overline{\sigma}^{(m,n,t)L}(g) < +\infty$ where $m-1 \leq q$. Also let f_u and g_v be integer translations of $f(z)$ and $g(z)$, respectively, for $u, v \in \mathbb{N}$. Then*

$$\liminf_{r\to+\infty} \frac{\log^{[p]} T(r, f_u(g_v))}{\log^{[m]} T(\exp^{[n]}[\log^{[n-1]} r \cdot \exp^{[t+1]} L(r)]^{\rho^{(m,n,t)L}(g)}), g_v)} \leq \frac{\rho^{(p,q,t)L}(f) \cdot \overline{\sigma}^{(m,n,t)L}(g)}{\lambda^{(m,n,t)L}(g)},$$

when $\exp^{[t]} L(M(r,g)) = o([\log^{[n-1]} r \cdot \exp^{[t+1]} L(r)]^{\alpha})$ as $r \to +\infty$ and for some positive $\alpha < \rho^{(m,n,t)L}(g)$.

Theorem 5.3.18 *Let $f(z)$ be a meromorphic function and $g(z)$ be a non constant entire function such that $\lambda^{(m,n,t)L}(g) > 0$, $\rho^{(p,q,t)L}(f) < +\infty$ and $\overline{\tau}^{(m,n,t)L}(g) < +\infty\infty$ where $m-1 \leq q$. Also let f_u and g_v be integer translations of $f(z)$ and $g(z)$, respectively, for $u, v \in \mathbb{N}$. Then*

$$\liminf_{r\to+\infty} \frac{\log^{[p]} T(r, f_u(g_v))}{\log^{[m]} T(\exp^{[n]}[\log^{[n-1]} r \cdot \exp^{[t+1]} L(r)]^{\lambda^{(m,n,t)L}(g)}), g_v)} \leq \frac{\rho^{(p,q,t)L}(f) \cdot \overline{\tau}^{(m,n,t)L}(g)}{\lambda^{(m,n,t)L}(g)},$$

when $\exp^{[t]} L(M(r,g)) = o([\log^{[n-1]} r \cdot \exp^{[t+1]} L(r)]^{\alpha})$ as $r \to +\infty$ and for some positive $\alpha < \lambda^{(m,n,t)L}(g)$.

Theorem 5.3.19 *Let $f(z)$ be a meromorphic function and $g(z)$ be a non constant entire function such that $\lambda^{(m,n,t)L}(g) > 0$, $\lambda^{(p,q,t)L}(f) < +\infty$ and $\tau^{(m,n,t)L}(g) < +\infty$ where $m-1 \leq q$. Also let f_u and g_v be integer translations of $f(z)$ and $g(z)$, respectively, for $u, v \in \mathbb{N}$. Then*

$$\liminf_{r\to+\infty} \frac{\log^{[p]} T(r, f_u(g_v))}{\log^{[m]} T(\exp^{[n]}[\log^{[n-1]} r \cdot \exp^{[t+1]} L(r)]^{\lambda^{(m,n,t)L}(g)}), g_v)} \leq \frac{\lambda^{(p,q,t)L}(f) \cdot \tau^{(m,n,t)L}(g)}{\lambda^{(m,n,t)L}(g)},$$

when $\exp^{[t]} L(M(r,g)) = o([\log^{[n-1]} r \cdot \exp^{[t+1]} L(r)]^{\alpha})$ *as* $r \to +\infty$ *and for some positive* $\alpha < \lambda^{(m,n,t)L}(g)$.

Theorem 5.3.20 *Let $f(z)$ be a meromorphic function and $g(z)$ be a non constant entire function such that $\rho^{(m,n,t)L}(g) > 0$, $\rho^{(p,q,t)L}(f) < +\infty$ and $\tau^{(m,n,t)L}(g) < +\infty$ where $m-1 \leq q$. Also let f_u and g_v be integer translations of $f(z)$ and $g(z)$, respectively, for $u, v \in \mathbb{N}$. Then*

$$\liminf_{r\to+\infty} \frac{\log^{[p]} T(r, f_u(g_v))}{\log^{[m]} T(\exp^{[n]}[\log^{[n-1]} r \cdot \exp^{[t+1]} L(r)]^{\lambda^{(m,n,t)L}(g)}), g_v)} \leq \frac{\rho^{(p,q,t)L}(f) \cdot \tau^{(m,n,t)L}(g)}{\rho^{(m,n,t)L}(g)},$$

when $\exp^{[t]} L(M(r,g)) = o([\log^{[n-1]} r \cdot \exp^{[t+1]} L(r)]^{\alpha})$ *as* $r \to +\infty$ *and for some positive* $\alpha < \lambda^{(m,n,t)L}(g)$.

Theorem 5.3.21 *Let $f(z)$ be a meromorphic function and $g(z)$ be a non constant entire function such that $\lambda^{(m,n,t)L}(g) > 0$, $\rho^{(p,q,t)L}(f) < +\infty$ and $\tau^{(m,n,t)L}(g) < +\infty$ where $m-1 \leq q$. Also let f_u and g_v be integer translations of $f(z)$ and $g(z)$, respectively, for $u, v \in \mathbb{N}$. Then*

$$\limsup_{r\to+\infty} \frac{\log^{[p]} T(r, f_u(g_v))}{\log^{[m]} T(\exp^{[n]}[\log^{[n-1]} r \cdot \exp^{[t+1]} L(r)]^{\lambda^{(m,n,t)L}(g)}), g_v)} \leq \frac{\rho^{(p,q,t)L}(f) \cdot \tau^{(m,n,t)L}(g)}{\lambda^{(m,n,t)L}(g)},$$

when $\exp^{[t]} L(M(r,g)) = o([\log^{[n-1]} r \cdot \exp^{[t+1]} L(r)]^{\alpha})$ *as* $r \to +\infty$ *and for some positive* $\alpha < \lambda^{(m,n,t)L}(g)$.

5.4 Concluding Remark

The main aim of this chapter is actually to extend and to modify the notion of order and lower order to $(p, q, t)L$-th order and $(p, q, t)L$-th lower order in case of integer translated composite entire and meromorphic functions and apply to establish some related growth properties.

However, the results of this chapter are basically inclined to prove some interesting results. But we also may have a scope to establish some theorems in order to get more illuminating results. Keeping this in mind, the theories, as well as the results of the next chapter, have been tackled under some different conditions.

References

[1] W. K. Hayman, *Meromorphic Functions*, Oxford: The Clarendon Press, 1964.

[2] T.V. Lakshminarasimhan, "A note on entire functions of bounded index", *J. Indian Math. Soc.*, Vol. 38, pp. 43-49, 1974.

[3] I. Lahiri and N.R. Bhattacharjee, "Functions of L-bounded index and of non-uniform L-bounded index", *Indian J. Math.*, Vol. 40, No. 1, pp. 43-57, 1998.

[4] T. Biswas, "Advancement on the study of growth analysis of differential polynomial and differential monomial in the light of slowly increasing functions", *Carpathian Math. Publ.*, Vol. 10, pp. 31-57, 2018.

[5] T. Biswas, "Comparative growth measurement of differential monomials and differential polynomials depending upon their relative ${}_pL^*$-types and relative ${}_pL^*$-weak types" *Aligarh Bull. Math.*, Vol. 36, pp. 73-94, 2017.

[6] T. Biswas, "Comparative growth analysis of differential monomials and differential polynomials depending on their relative ${}_pL^*$-orders", *J. Chungcheong Math. Soc.*, Vol.31, pp. 103-130, 2018.

[7] W. Bergweiler, "On the Nevanlinna Characteristic of a composite function", *Complex Variables*, Vol. 10, pp.225-236, 1988.

CHAPTER 6

Some Growth Properties of Integer Translated Composite Entire and Meromorphic Functions on the Basis of $(p, q, t)L$-th Order and $(p, q, t)L$-th Type

fancyhead

Abstract: In this chapter, we establish some new results depending on the comparative growth properties of the composition of integer translated entire and meromorphic functions using $(p, q, t)L$-th order and $(p, q, t)L$-th type.

Keywords: Integer translated entire function, Integer translated meromorphic function, $(p, q, t)L$-th order, $(p, q, t)L$-th type.
Mathematics Subject Classification (2020) : 30D30, 30D35.

6.1 Introduction

Lakshminarasimhan [1] introduced the idea of the functions of the L-bounded index. Later Lahiri and Bhattacharjee [2] worked on the entire functions of the L-bounded index and of the non uniform L-bounded index. In this Chapter we establish some new results depending on the comparative growth properties of composition of integer translated entire or meromorphic functions using $(p, q, t)L$-th order, $(p, q, t)L$-th type and $(p, q, t)L$-th weak type.

6.2 Lemmas

In this section, we present some lemmas which will be needed in the sequel.

Tanmay Biswas and Chinmay Biswas

Lemma 6.2.1 [3] *If $f(z)$ be meromorphic and $g(z)$ be entire then for all sufficiently large values of r,*

$$T(r, f(g)) \leq \{1 + o(1)\} \frac{T(r, g)}{\log M(r, g)} T(M(r, g), f).$$

6.3 Main Results

In this section, we present the main results of the chapter.

Theorem 6.3.1 *Let $f(z)$ be a meromorphic function and $g(z)$ be a non constant entire function such that $\rho^{(p,q,t)L}(f) = \rho^{(m,n,t)L}(g)$, $0 < \sigma^{(m,n,t)L}(g) < +\infty$ and $\overline{\sigma}^{(p,q,t)L}(f) > 0$ where $m - 1 = n = q$ and $p > 2$. Also let f_u and g_v be integer translations of $f(z)$ and $g(z)$, respectively, for $u, v \in \mathbb{N}$. Then*

$$\limsup_{r \to +\infty} \frac{\log^{[p]} T(r, f_u(g_v))}{\log^{[p-1]} T(r, f_u) + \exp^{[t]} L(M(r, g))}$$
$$\leq \begin{cases} \frac{\rho^{(p,q,t)L}(f)\sigma^{(m,n,t)L}(g)}{\overline{\sigma}^{(p,q,t)L}(f)} & \text{if } \exp^{[t]} L(M(r, g)) = o\{\log^{[p-1]} T(r, f_u)\} \\ \rho^{(p,q,t)L}(f) & \text{if } \log^{[p-1]} T(r, f_u) = o\{\exp^{[t]} L(M(r, g))\}. \end{cases}.$$

Proof Let $f_u \circ g_v = h_t$, where is a meromorphic function and $t \in \mathbb{N}$. So h_t can be expressed as

$$h_t = \{(z + t) : t \in \mathbb{N}\}.$$

Then by (2.1.3) we obtain

$$T(r, h_t) = tT(r, h) + \sum_t (e_t + e_t'),$$

where e_t, $e_t' \to 0$ as $r \to +\infty$,

$$i.e., T(r, f_u \circ g_v) = tT(r, f \circ g) + \sum_t (e_t + e_t'). \tag{6.3.1}$$

Now in view of Lemma 6.2.1 and the inequality $T(r, g) \leq \log M(r, g)$ {cf. [4] } we get from (6.3.1) for all sufficiently large values of r

$$i.e., \ \log^{[p]} T(r, f_u(g_v)) \leqslant \log^{[p]} T_f(M(r, g)) + O(1) \tag{6.3.2}$$

$$i.e., \ \log^{[p]} T(r, f_u(g_v)) \leq$$
$$(\rho^{(p,q,t)L}(f) + \varepsilon)[\log^{[q]} M(r, g) + \exp^{[t]} L(M(r, g))] + O(1)$$
$$i.e., \ \log^{[p]} T(r, f_u(g_v)) \leq (\rho^{(p,q,t)L}(f) + \varepsilon)[\log^{[m-1]} M(r, g) + \exp^{[t]} L(M(r, g))] + O(1)$$
$$i.e., \ \log^{[p]} T(r, f_u(g_v)) \leq (\rho^{(p,q,t)L}(f) + \varepsilon) \cdot$$

$$[(\sigma^{(m,n,t)L}(g)+\varepsilon)[\log^{[n-1]} r\cdot \exp^{[t+1]} L(r)]^{\rho^{(m,n,t)L}(g)}+\exp^{[t]} L(M(r,g))]+O(1).$$

Since $\rho^{(p,q,t)L}(f)=\rho^{(m,n,t)L}(g)$, we obtain from above for all sufficiently large values of r

$$i.e.,\ \log^{[p]} T(r,f_u(g_v))\leq (\rho^{(p,q,t)L}(f)+\varepsilon)\cdot$$

$$[(\sigma^{(m,n,t)L}(g)+\varepsilon)[\log^{[n-1]} r\cdot \exp^{[t+1]} L(r)]^{\rho^{(p,q,t)L}(f)}+\exp^{[t]} L(M(r,g))]+O(1). \quad (6.3.3)$$

Again in view of Lemma 2.2.2, Lemma 4.2.2, we get for all sufficiently large values of r,

$$\log^{[p-1]} T(r,f_u)\geq (\overline{\sigma}^{(p,q,t)L}(f_u)-\varepsilon)[\log^{[q-1]} r\cdot \exp^{[t+1]} L(r)]^{\rho^{(p,q,t)L}(f_u)}$$

$$i.e.,\ \log^{[p-1]} T(r,f_u)\geq (\overline{\sigma}^{(p,q,t)L}(f)-\varepsilon)[\log^{[q-1]} r\cdot \exp^{[t+1]} L(r)]^{\rho^{(p,q,t)L}(f)}$$

$$i.e.,\ [\log^{[q-1]} r\cdot \exp^{[t+1]} L(r)]^{\rho^{(p,q,t)L}(f)}\leq \frac{\log^{[p-1]} T(r,f_u)}{(\overline{\sigma}^{(p,q,t)L}(f)-\varepsilon)}$$

$$i.e.,\ [\log^{[n-1]} r\cdot \exp^{[t+1]} L(r)]^{\rho^{(p,q,t)L}(f)}\leq \frac{\log^{[p-1]} T(r,f_u)}{(\overline{\sigma}^{(p,q,t)L}(f)-\varepsilon)}. \quad (6.3.4)$$

Now from (6.3.3) and (6.3.4) it follows for all sufficiently large values of r

$$\log^{[p]} T(r,f_u(g_v))\leq (\rho^{(p,q,t)L}(f)+\varepsilon)\cdot \exp^{[t]} L(M(r,g))+O(1)+$$

$$(\rho^{(p,q,t)L}(f)+\varepsilon)(\sigma^{(m,n,t)L}(g)+\varepsilon)\cdot \frac{\log^{[p-1]} T(r,f_u)}{(\overline{\sigma}^{(p,q,t)L}(f)-\varepsilon)}$$

$$ie.,\ \frac{\log^{[p]} T(r,f_u(g_v))}{\log^{[p-1]} T(r,f_u)+\exp^{[t]} L(M(r,g))}\leq \frac{O(1)}{\log^{[p-1]} T(r,f_u)+\exp^{[t]} L(M(r,g))}$$

$$+\frac{(\rho^{(p,q,t)L}(f)+\varepsilon)}{1+\frac{\log^{[p-1]} T(r,f_u)}{\exp^{[t]} L(M(r,g))}}+\frac{\frac{(\rho^{(p,q,t)L}(f)+\varepsilon)(\sigma^{(m,n,t)L}(g)+\varepsilon)}{(\overline{\sigma}^{(p,q,t)L}(f)-\varepsilon)}}{1+\frac{\exp^{[t]} L(M(r,g))}{\log^{[p-1]} T(r,f_u)}}. \quad (6.3.5)$$

If $\exp^{[t]} L(M(r,g))=o\{\log^{[p-1]} T(r,f_u)\}$ then from (6.3.5) we get

$$\limsup_{r\to+\infty}\frac{\log^{[p]} T(r,f_u(g_v))}{\log^{[p-1]} T(r,f_u)+\exp^{[t]} L(M(r,g))}\leq \frac{(\rho^{(p,q,t)L}(f)+\varepsilon)(\sigma^{(m,n,t)L}(g)+\varepsilon)}{(\overline{\sigma}^{(p,q,t)L}(f)-\varepsilon)}.$$

Since $\varepsilon(>0)$ is arbitrary, it follows from above

$$\limsup_{r\to+\infty}\frac{\log^{[p]} T(r,f_u(g_v))}{\log^{[p-1]} T(r,f_u)+\exp^{[t]} L(M(r,g))}\leq \frac{\rho^{(p,q,t)L}(f)\sigma^{(m,n,t)L}(g)}{\overline{\sigma}^{(p,q,t)L}(f)}$$

Again if $\log^{[p-1]} T(r,f_u)=o\{\exp^{[t]} L(M(r,g))\}$ then from (6.3.5) it follows

$$\limsup_{r\to+\infty}\frac{\log^{[p]} T(r,f_u(g_v))}{\log^{[p-1]} T(r,f_u)+\exp^{[t]} L(M(r,g))}\leq (\rho^{(p,q,t)L}(f)+\varepsilon).$$

As $\varepsilon(>0)$ is arbitrary, we obtain from above

$$\limsup_{r\to+\infty}\frac{\log^{[p]} T(r,f_u(g_v))}{\log^{[p-1]} T(r,f_u)+\exp^{[t]} L(M(r,g))}\leq \rho^{(p,q,t)L}(f).$$

Thus the theorem is established.

Theorem 6.3.2 *Let $f(z)$ be a meromorphic function and $g(z)$ be a non constant entire function such that $\lambda^{(p,q,t)L}(f) < 0$, $\rho^{(p,q,t)L}(f) = \rho^{(m,n,t)L}(g)$, $0 < \sigma^{(m,n,t)L}(g) < +\infty$ and $\overline{\sigma}^{(p,q,t)L}(f) > 0$ where $m-1=n=q$ and $p>2$. Also let f_u and g_v be integer translations of $f(z)$ and $g(z)$, respectively, for $u, v \in \mathbb{N}$. Then*

$$\underline{\lim}_{r\to+\infty} \frac{\log^{[p]} T(r, f_u(g_v))}{\log^{[p-1]} T(r, f_u) + \exp^{[t]} L(M(r,g))} \le \begin{cases} \frac{\lambda^{(p,q,t)L}(f)\sigma^{(m,n,t)L}(g)}{\overline{\sigma}^{(p,q,t)L}(f)} & \text{if } \exp^{[t]} L(M(r,g)) = o\{\log^{[p-1]} T(r, f_u)\} \\ \lambda^{(p,q,t)L}(f) & \text{if } \log^{[p-1]} T(r, f_u) = o\{\exp^{[t]} L(M(r,g))\}. \end{cases}$$

Theorem 6.3.3 *Let $f(z)$ be a meromorphic function and $g(z)$ be a non constant entire function such that $\rho^{(p,q,t)L}(f) = \rho^{(m,n,t)L}(g)$, $0 < \sigma^{(m,n,t)L}(g) < +\infty$ and $\sigma^{(p,q,t)L}(f) > 0$ where $m-1=n=q$ and $p>2$. Also let f_u and g_v be integer translations of $f(z)$ and $g(z)$, respectively, for $u, v \in \mathbb{N}$. Then*

$$\underline{\lim}_{r\to+\infty} \frac{\log^{[p]} T(r, f_u(g_v))}{\log^{[p-1]} T(r, f_u) + \exp^{[t]} L(M(r,g))} \le \begin{cases} \frac{\rho^{(p,q,t)L}(f)\sigma^{(m,n,t)L}(g)}{\sigma^{(p,q,t)L}(f)} & \text{if } \exp^{[t]} L(M(r,g)) = o\{\log^{[p-1]} T(r, f_u)\} \\ \rho^{(p,q,t)L}(f) & \text{if } \log^{[p-1]} T(r, f_u) = o\{\exp^{[t]} L(M(r,g))\}. \end{cases}$$

Theorem 6.3.4 *Let $f(z)$ be a meromorphic function and $g(z)$ be a non constant entire function such that $\rho^{(p,q,t)L}(f) = \rho^{(m,n,t)L}(g)$, $0 < \overline{\sigma}^{(m,n,t)L}(g) < +\infty$ and $\overline{\sigma}^{(p,q,t)L}(f) > 0$ where $m-1=n=q$ and $p>2$. Also let f_u and g_v be integer translations of $f(z)$ and $g(z)$, respectively, for $u, v \in \mathbb{N}$. Then*

$$\underline{\lim}_{r\to+\infty} \frac{\log^{[p]} T(r, f_u(g_v))}{\log^{[p-1]} T(r, f_u) + \exp^{[t]} L(M(r,g))} \le \begin{cases} \frac{\rho^{(p,q,t)L}(f)\overline{\sigma}^{(m,n,t)L}(g)}{\overline{\sigma}^{(p,q,t)L}(f)} & \text{if } \exp^{[t]} L(M(r,g)) = o\{\log^{[p-1]} T(r, f_u)\} \\ \rho^{(p,q,t)L}(f) & \text{if } \log^{[p-1]} T(r, f_u) = o\{\exp^{[t]} L(M(r,g))\}. \end{cases}$$

We omit the proof of the above three theorems as those can be carried out in the line of Theorem 6.3.1.

Similarly, using the concept of the growth indicator $\overline{\tau}^{(p,q,t)L}(f)$ and $\tau^{(m,n,t)L}(g)$, we may state the subsequent four theorems without their proofs since those can be carried out in the line of Theorem 6.3.1, Theorem 6.3.2, Theorem 6.3.3 and Theorem 6.3.4 respectively and with the help of Lemma 2.2.2 and Lemma 4.2.2.

Theorem 6.3.5 *Let $f(z)$ be a meromorphic function and $g(z)$ be a non constant entire function such that $\rho^{(p,q,t)L}(f) < +\infty$, $\lambda^{(p,q,t)L}(f) = \lambda^{(m,n,t)L}(g)$, $0 < \tau^{(m,n,t)L}(g) < +\infty$ and*

$\overline{\tau}^{(p,q,t)L}(f) > 0$ *where* $m - 1 = n = q$ *and* $p > 2$. *Also let* f_u *and* g_v *be integer translations of* $f(z)$ *and* $g(z)$, *respectively, for* $u, v \in \mathbb{N}$. *Then*

$$\limsup_{r\to+\infty} \frac{\log^{[p]} T(r, f_u(g_v))}{\log^{[p-1]} T(r, f_u) + \exp^{[t]} L(M(r,g))} \le \begin{cases} \frac{\rho^{(p,q,t)L}(f)\tau^{(m,n,t)L}(g)}{\overline{\tau}^{(p,q,t)L}(f)} & \text{if } \exp^{[t]} L(M(r,g)) = o\{\log^{[p-1]} T(r, f_u)\} \\ \rho^{(p,q,t)L}(f) & \text{if } \log^{[p-1]} T(r, f_u) = o\{\exp^{[t]} L(M(r,g))\}. \end{cases}.$$

Theorem 6.3.6 *Let* $f(z)$ *be a meromorphic function and* $g(z)$ *be a non constant entire function such that* $\lambda^{(p,q,t)L}(f) = \lambda^{(m,n,t)L}(g)$, $0 < \tau^{(m,n,t)L}(g) < +\infty$ *and* $\overline{\tau}^{(p,q,t)L}(f) > 0$ *where* $m - 1 = n = q$ *and* $p > 2$. *Also let* f_u *and* g_v *be integer translations of* $f(z)$ *and* $g(z)$, *respectively, for* $u, v \in \mathbb{N}$. *Then*

$$\liminf_{r\to+\infty} \frac{\log^{[p]} T(r, f_u(g_v))}{\log^{[p-1]} T(r, f_u) + \exp^{[t]} L(M(r,g))} \le \begin{cases} \frac{\lambda^{(p,q,t)L}(f)\tau^{(m,n,t)L}(g)}{\overline{\tau}^{(p,q,t)L}(f)} & \text{if } \exp^{[t]} L(M(r,g)) = o\{\log^{[p-1]} T(r, f_u)\} \\ \lambda^{(p,q,t)L}(f) & \text{if } \log^{[p-1]} T(r, f_u) = o\{\exp^{[t]} L(M(r,g))\}. \end{cases}.$$

Theorem 6.3.7 *Let* $f(z)$ *be a meromorphic function and* $g(z)$ *be a non constant entire function such that* $\rho^{(p,q,t)L}(f) < +\infty$, $\lambda^{(p,q,t)L}(f) = \lambda^{(m,n,t)L}(g)$, $0 < \tau^{(m,n,t)L}(g) < +\infty$ *and* $\tau^{(p,q,t)L}(f) > 0$ *where* $m - 1 = n = q$ *and* $p > 2$. *Also let* f_u *and* g_v *be integer translations of* $f(z)$ *and* $g(z)$, *respectively, for* $u, v \in \mathbb{N}$. *Then*

$$\liminf_{r\to+\infty} \frac{\log^{[p]} T(r, f_u(g_v))}{\log^{[p-1]} T(r, f_u) + \exp^{[t]} L(M(r,g))} \le \begin{cases} \frac{\rho^{(p,q,t)L}(f)\tau^{(m,n,t)L}(g)}{\tau^{(p,q,t)L}(f)} & \text{if } \exp^{[t]} L(M(r,g)) = o\{\log^{[p-1]} T(r, f_u)\} \\ \rho^{(p,q,t)L}(f) & \text{if } \log^{[p-1]} T(r, f_u) = o\{\exp^{[t]} L(M(r,g))\}. \end{cases}.$$

Theorem 6.3.8 *Let* $f(z)$ *be a meromorphic function and* $g(z)$ *be a non constant entire function such that* $\rho^{(p,q,t)L}(f) < +\infty$, $\lambda^{(p,q,t)L}(f) = \lambda^{(m,n,t)L}(g)$, $0 < \overline{\tau}^{(m,n,t)L}(g) < +\infty$ *and* $\overline{\tau}^{(p,q,t)L}(f) > 0$ *where* $m - 1 = n = q$ *and* $p > 2$. *Also let* f_u *and* g_v *be integer translations of* $f(z)$ *and* $g(z)$, *respectively, for* $u, v \in \mathbb{N}$. *Then*

$$\liminf_{r\to+\infty} \frac{\log^{[p]} T(r, f_u(g_v))}{\log^{[p-1]} T(r, f_u) + \exp^{[t]} L(M(r,g))} \le \begin{cases} \frac{\rho^{(p,q,t)L}(f)\overline{\tau}^{(m,n,t)L}(g)}{\overline{\tau}^{(p,q,t)L}(f)} & \text{if } \exp^{[t]} L(M(r,g)) = o\{\log^{[p-1]} T(r, f_u)\} \\ \rho^{(p,q,t)L}(f) & \text{if } \log^{[p-1]} T(r, f_u) = o\{\exp^{[t]} L(M(r,g))\}. \end{cases}.$$

Analogously we state the following four theorems under some different conditions, which can also be carried out using the same technique of Theorem 6.3.1 and with the help of Lemma 2.2.2 and Lemma 4.2.2, respectively. Hence their proofs are omitted.

Theorem 6.3.9 *Let $f(z)$ be a meromorphic function and $g(z)$ be a non constant entire function such that $\rho^{(p,q,t)L}(f) < +\infty$, $\lambda^{(p,q,t)L}(f) = \rho^{(m,n,t)L}(g)$, $0 < \sigma^{(m,n,t)L}(g) < +\infty$ and $\overline{\tau}^{(p,q,t)L}(f) > 0$ where $m-1 = n = q$ and $p > 2$. Also let f_u and g_v be integer translations of $f(z)$ and $g(z)$, respectively, for $u, v \in \mathbb{N}$. Then*

$$\limsup_{r\to+\infty} \frac{\log^{[p]} T(r, f_u(g_v))}{\log^{[p-1]} T(r, f_u) + \exp^{[t]} L(M(r,g))} \leq \begin{cases} \frac{\rho^{(p,q,t)L}(f)\sigma^{(m,n,t)L}(g)}{\overline{\tau}^{(p,q,t)L}(f)} & \text{if } \exp^{[t]} L(M(r,g)) = o\{\log^{[p-1]} T(r, f_u)\} \\ \rho^{(p,q,t)L}(f) & \text{if } \log^{[p-1]} T(r, f_u) = o\{\exp^{[t]} L(M(r,g))\}. \end{cases}.$$

Theorem 6.3.10 *Let $f(z)$ be a meromorphic function and $g(z)$ be a non constant entire function such that $\lambda^{(p,q,t)L}(f) = \rho^{(m,n,t)L}(g)$, $0 < \sigma^{(m,n,t)L}(g) < +\infty$ and $\overline{\tau}^{(p,q,t)L}(f) > 0$ where $m-1 = n = q$ and $p > 2$. Also let f_u and g_v be integer translations of $f(z)$ and $g(z)$, respectively, for $u, v \in \mathbb{N}$. Then*

$$\lim_{r\to+\infty} \frac{\log^{[p]} T(r, f_u(g_v))}{\log^{[p-1]} T(r, f_u) + \exp^{[t]} L(M(r,g))} \leq \begin{cases} \frac{\lambda^{(p,q,t)L}(f)\sigma^{(m,n,t)L}(g)}{\overline{\tau}^{(p,q,t)L}(f)} & \text{if } \exp^{[t]} L(M(r,g)) = o\{\log^{[p-1]} T(r, f_u)\} \\ \lambda^{(p,q,t)L}(f) & \text{if } \log^{[p-1]} T(r, f_u) = o\{\exp^{[t]} L(M(r,g))\}. \end{cases}.$$

Theorem 6.3.11 *Let $f(z)$ be a meromorphic function and $g(z)$ be a non constant entire function such that $\rho^{(p,q,t)L}(f) = \lambda^{(m,n,t)L}(g)$, $0 < \tau^{(m,n,t)L}(g) < +\infty$ and $\overline{\sigma}^{(p,q,t)L}(f) > 0$ where $m-1 = n = q$ and $p > 2$. Also let f_u and g_v be integer translations of $f(z)$ and $g(z)$, respectively, for $u, v \in \mathbb{N}$. Then*

$$\limsup_{r\to+\infty} \frac{\log^{[p]} T(r, f_u(g_v))}{\log^{[p-1]} T(r, f_u) + \exp^{[t]} L(M(r,g))} \leq \begin{cases} \frac{\rho^{(p,q,t)L}(f)\tau^{(m,n,t)L}(g)}{\overline{\sigma}^{(p,q,t)L}(f)} & \text{if } \exp^{[t]} L(M(r,g)) = o\{\log^{[p-1]} T(r, f_u)\} \\ \rho^{(p,q,t)L}(f) & \text{if } \log^{[p-1]} T(r, f_u) = o\{\exp^{[t]} L(M(r,g))\}. \end{cases}.$$

Theorem 6.3.12 *Let $f(z)$ be a meromorphic function and $g(z)$ be a non constant entire function such that $\rho^{(p,q,t)L}(f) = \lambda^{(m,n,t)L}(g)$, $0 < \tau^{(m,n,t)L}(g) < +\infty$ and $\overline{\sigma}^{(p,q,t)L}(f) > 0$ where $m-1 = n = q$ and $p > 2$. Also let f_u and g_v be integer translations of $f(z)$ and*

$g(z)$, *respectively, for* $u, v \in \mathbb{N}$. *Then*

$$\varlimsup_{r \to +\infty} \frac{\log^{[p]} T(r, f_u(g_v))}{\log^{[p-1]} T(r, f_u) + \exp^{[t]} L(M(r,g))} \leq \begin{cases} \frac{\lambda^{(p,q,t)L}(f) \tau^{(m,n,t)L}(g)}{\overline{\sigma}^{(p,q,t)L}(f)} & \text{if } \exp^{[t]} L(M(r,g)) = o\{\log^{[p-1]} T(r, f_u)\} \\ \lambda^{(p,q,t)L}(f) & \text{if } \log^{[p-1]} T(r, f_u) = o\{\exp^{[t]} L(M(r,g))\}. \end{cases}$$

6.4 Concluding Remark

The main aim of this chapter is actually to apply the notion of $(p,q,t)L$-th order and $(p,q,t)L$-th type for estimating the growth properties of compositing of integer translated entire and meromorphic functions under some different conditions. Moreover, the notion of $(p,q,t)L$-th order and $(p,q,t)L$-th type of higher dimensions may also be applied in case of entire or meromorphic functions of several complex variables.

References

[1] T.V. Lakshminarasimhan, "A note on entire functions of bounded index", *J. Indian Math. Soc.*, Vol. 38, pp. 43-49, 1974.

[2] I. Lahiri and N.R. Bhattacharjee, "Functions of L-bounded index and of non-uniform L-bounded index", *Indian J. Math.*, Vol. 40, No. 1, pp. 43-57, 1998.

[3] W. Bergweiler, "On the Nevanlinna Characteristic of a composite function, Complex Variables", Vol. 10, pp.225-236, 1988.

[4] W. K. Hayman, *Meromorphic Functions*, Oxford: The Clarendon Press, 1964.

SUBJECT INDEX

S

www.ingramcontent.com/pod-product-compliance
Lightning Source LLC
Chambersburg PA
CBHW041723210326
41598CB00007B/753